1.7.45œ   22⁵⁰

W.P. protection —————— 62
→ Coo... ——————— 23 +
  ... ———————————— 64
  ... roof membrane
  (Protected ") — 25 +
  Gyp. Bd underlayment ————— 80
  Roof vents ————————— 26
  Compact roof ————————— 24

Repairing and Extending Weather Barriers

VAN NOSTRAND REINHOLD'S
BUILDING RENOVATION AND RESTORATION SERIES

Repairing and Extending Weather Barriers   ISBN 0-442-20611-9

Repairing and Extending Finishes, Part I   ISBN 0-442-20612-7

Repairing and Extending Finishes, Part II   ISBN 0-442-20613-5

Repairing and Extending Non-Structural Metals   ISBN 0-442-20615-1

Repairing and Extending Doors, Windows, and Cladding   ISBN 0-442-20618-6

Repairing, Extending, and Cleaning Brick and Block   ISBN 0-442-20619-4

Repairing, Extending, and Cleaning Stone   ISBN 0-442-20620-8

Repairing and Extending Wood   ISBN 0-442-20621-6

Building Renovation and Restoration Series

Repairing
and Extending
Weather Barriers

H. Leslie Simmons, AIA, CSI

VNR VAN NOSTRAND REINHOLD
New York

Printed in the United States of America

Designed by Caliber Design

Van Nostrand Reinhold
115 Fifth Avenue
New York, New York 10003

Van Nostrand Reinhold International Company Limited
11 New Fetter Lane
London EC4P 4EE, England

Van Nostrand Reinhold
480 La Trobe Street
Melbourne, Victoria 3000, Australia

Macmillan of Canada
Division of Canada Publishing Corporation
164 Commander Boulevard
Agincourt, Ontario M1S 3C7, Canada

16 15 14 13 12 11 10 9 8 7 6 5 4 3 2 1

Library of Congress Cataloging in Publication Data

Simmons, H. Leslie.
  Repairing and extending weather barriers / by H. Leslie Simmons.
    p. cm.—(Building renovation and restoration ; v. 1)
  Bibliography: p.
  Includes index.
  ISBN 0-442-20611-9
  1. Buildings—Protection. 2. Dampness in buildings.
3. Insulation (Heat) 4. Building materials. I. Title.
II. Series.
TH9025.S525   1989
693.8—dc19   89-5503

Foreword

To spite a national trend toward renovation, restoration, and remodeling, construction products producers and their associations are not universally eager to publish recommendations for repairing or extending existing materials. There are two major reasons. First, there are several possible applications of most building materials; and there is an even larger number of different problems that can occur after products are installed in a building. Thus, it is difficult to produce recommendations that cover every eventuality.

Second, it is not always in a building construction product producer's best interest to publish data that will help building owners repair their product. Producers, whose income derives from selling new products, do not necessarily applaud when their associations spend their money telling architects and building owners how to avoid buying their products.

Finally, in the *Building Renovation and Restoration Series* we have a reference that recognizes that problems frequently occur with materials used in building projects. In this book and in the other books in this series,

Simmons goes beyond the promotional hyperbole found in most product literature and explains how to identify common problems. He then offers informed "inside" recommendations on how to deal with each of the problems. Each chapter covers certain materials, or family of materials, in a way that can be understood by building owners and managers, as well as construction and design professionals.

Most people involved in designing, financing, constructing, owning, managing, and maintaining today's "high tech" buildings have limited knowledge on how all of the many materials go together to form the building and how they should look and perform. Everyone relies on specialists, who may have varying degrees of expertise, for building and installing the many individual components that make up completed buildings. Problems frequently arise when components and materials are not installed properly and often occur when substrate or supporting materials are not installed correctly.

When problems occur, even the specialists may not know why they are happening. Or they may not be willing to admit responsibility for problems. Such problems can stem from improper designer selection, defective or substandard installation, lack of understanding on use, or incorrect maintenance procedures. Armed with necessary "inside" information, one can identify causes of problems and make assessments of their extent. Only after the causes are identified can one determine how to correct the problem.

Up until now that "inside" information generally was not available to those faced with these problems. In this book, materials are described according to types and uses and how they are supposed to be installed or applied. Materials and installation or application failures and problems are identified and listed, then described in straightforward, understandable language supplemented by charts, graphs, photographs, and line drawings. Solutions ranging from proper cleaning and other maintenance and remedial repair to complete removal and replacement are recommended with cross-references to given problems.

Of further value are sections on where to get more information from such sources as manufacturers, standards setting bodies, government agencies, periodicals, and books. There are also national and regional trade and professional associations representing almost every building and finish material, most of which make available reliable, unbiased information on proper use and installation of their respective materials and products. Some associations even offer information on recognizing and solving problems for their products and materials. Names, addresses, and telephone numbers are included, along with each association's major publications. In addition, knowledgeable, independent consultants who specialize in resolving problems relating to certain materials are recognized. Where names were not available for publication, most associations can furnish names of qualified persons who can assist in resolving problems related to their products.

It is wished that one would never be faced with any problems with new buildings and even older ones. However, reality being what it is, this book, as do the others in the series, offers a guide so you can identify problems and find solutions. And it provides references for sources of more information when problems go beyond the scope of the book.

Jess McIlvain, AIA, CCS, CSI
Consulting Architect
Bethesda, Maryland

Contents

CHAPTER 2 Condensation Control   **18**

CHAPTER 3 Bituminous Dampproofing and Clear Water Repellents   **36**

CHAPTER 4 Waterproofing   **55**

CHAPTER 7   Low-Slope Roofing   195

CHAPTER 8   Flashing, Roof Specialties, and Sheet Metal Roofing   251

Preface

Architects working on projects where existing construction plays a part spend countless hours eliciting from materials producers and installers data relating to cleaning, repairing, and extending existing building materials and products and for installing new materials and products over existing materials. The producers and installers know much of the needed information and generally give it up readily when asked, but they often do not include such information in their standard literature packages. As a result, there is a long-standing need for source documents that included the industry's recommendations for repairing, maintaining, and extending existing materials and for installing new materials over existing materials. This book is one of a series called *Building Renovation and Restoration Series* that was conceived to answer that need.

In the thirty-plus years I have worked as an architect, and especially since 1975, when I began my practice as a specifications consultant, I have looked without success for a central source of data about weather barriers.

In recent years, publications by the National Roofing Contractors Association have become a major source of help, but they say little about, and sometimes do not agree with, what other industry experts recommend. The construction industry in general, and the roofing industry in particular, is fragmented and, sometimes, confrontational. There is no central source of data. Authoritative sources do not agree on many subjects. It is necessary for building owners, architects, consultants, and contractors to consult several sources to resolve many problems, which takes a great deal of research and time.

I have done much of that kind of research myself over the years. This book includes the fruits of my earlier research, augmented by many additional hours of recent searching to make the book as broad as possible. In it, I have included as many of the industry's recommendations on working with existing roofing, waterproofing, clear water repellents, dampproofing, and flashing as I could fit in. Since no single book could possibly contain all known data about those subjects, I have also identified and discussed other sources of data to help readers solve problems not fully discussed here.

With the publisher's concurrence, I decided to make this the first book in the series because the subjects discussed here are involved in the most prevalent problems faced by building owners, architects, engineers, and building contractors. Most buildings suffer water damage at some time during their life span. The damage may come from a number of sources. Plumbing and mechanical systems sometimes leak. Condensation sometimes occurs because of faulty design or construction of roofs or walls. But leaks probably are the largest contributor to water damage. In 1986, 80 percent of post-construction litigation involving new building construction resulted from roof leaks. There is no reason to believe the numbers are different today.

Any part of a building's shell may leak, of course, but most leaks come through barriers specifically designed to prevent leaks, including those barriers discussed in this book.

A major secondary problem associated with water leakage near the roof is wetting of the roof insulation. Wet insulation not only loses much of its insulating qualities but also can damage other roofing components and supports. Roof insulation is also discussed in this book.

The data in this book, as is true for that in the other books in the series as well, come from published recommendations of producers and their associations; applicable codes and standards; federal agency guides and requirements; contractors who actually do such work in the field; the experiences of other architects and their consultants; and from the author's own experiences. Where data are too voluminous to include in the text, references are given to help the reader find additional information from knowledgeable sources. Some sources of data about historic preservation are also listed.

This book, as do the others in the series, explains in practical, under-standable narrative, supported by line drawings, photographs, charts, and graphs, how to extend, clean, repair, refinish, restore, and protect the existing materials that are the subject of the book, and how to install the materials discussed in the book over existing materials.

All the books in the series are written for building owners; architects; federal and local government agencies; building contractors; university, professional, and public libraries; members of groups and associations in-terested in preservation; and everyone who is responsible for maintaining, cleaning, or repairing existing building construction materials. They are not how-to books meant to compete with publications such as *The Old House Journal* or the books and tapes generated by the producers of the television series, "This Old House."

I hope that if this book does not directly solve your current problem, it will lead you to a source that will.

How to Use This Book

This book is divided into eight chapters. Chapter 1 is a general introduction to the subject. It includes a discussion of methods an owner, architect, engineer, or building contractor can use to determine how to handle a problem associated with weather barriers, including the place of specialty consultants.

Chapters 2 through 8 discuss the subject areas suggested by their titles. They each include:

- A statement of the nature and purpose of that chapter
- A discussion of materials commonly used to form the type of barrier discussed in that chapter and their usual uses
- A discussion of how and why those materials fail
- Methods of extending and repairing those materials and installing them over existing materials
- An indication of sources of additional information about the subjects in that chapter

The book has an Appendix that contains a list of sources of additional data. Sources include manufacturers, trade and professional associations, standards setting bodies, government agencies, periodicals, book publishers, and others having knowledge of methods for restoring building materials. The list includes names, addresses, and telephone numbers. Sources from which data related to historic preservation may be obtained are identified with a boldface **HP**. Many of the publications and publishers of entries in the Bibliography are listed in the Appendix.

The items listed in the Bibliography are annotated to show the book chapters to which they apply. Entries that are related to historic preservation are identified with a boldface **HP**.

Building owners, engineers and architects, and general building contractors will, in most cases, use this book in a somewhat different way. The following suggestions give some indication of what some of those differences might be.

Owners

It is probably safe to assume that you are consulting this book because your building has experienced or is now experiencing a weather barrier failure. Most such failures are manifested by water leaking into your building. If the problem is an **EMERGENCY,** turn immediately to Chapter 1 and read the parts there called "What to Do in an Emergency," and in "Help for Owners" the part called "Emergencies."

When the failure is at least temporarily under control (water is not now leaking into your building), a more systematic approach is suggested. You will tend to want to turn directly to the chapter containing information about the barrier that seems to be leaking. If you have a good knowledge of such problems, and experience with them, you may be able to approach the problem in that manner. Otherwise it is better to first become familiar with the entire contents of this book. Begin by flipping through the contents pages and then read the first few paragraphs of each chapter, which describe what that chapter contains and its purpose.

After that, read all of Chapter 1, including those parts that do not at first seem to be applicable to your immediate problem. The source of a weather barrier failure is not always what it seems. Water apparently flowing through a roof may, for example, be a result of a flashing problem, or even condensation. Jumping to an incorrect conclusion can be costly.

After reading Chapter 1, then turn to the chapter covering the most probable source of your barrier failure. Chapter 1 offers advice to help you know where to look.

A word of caution, however. Unless you have extensive experience in dealing with such weather barrier failures, do not reach for the telephone to call for professional help until you have read the chapters covering the possible sources of failure. It is always better to know as much as possible about a problem before asking for help.

Architects and Engineers

Your approach will depend somewhat on your relationship to the owner. Suppose in the first instance that you have been called because an existing building has experienced a weather barrier failure. There will be obvious differences in your approach depending on whether you were the building's original designer, especially if the failure is within the normal expected life of the barrier that seems to have failed. There are, in that case, possible legal problems, as well as technical ones. This book, though, is limited to discussion of technical problems.

Your natural tendency may be to rush to the site to determine exactly what the problem is. That is not a bad idea if you have extensive experience with weather barrier failures. If you have little such experience, it is always better to do some homework before subjecting yourself to queries by a client or potential client.

That homework might consist of reading the chapters of this book that deal with the apparent problem. Then if the problem is even slightly beyond your expertise, read Chapter 1 and decide whether you might need outside professional help. If you have some experience, you might defer your decision about seeking professional help until you have had a chance to study the problem in the field. If you know little about the subject, you will probably want someone knowledgeable in the subject to accompany you on your first visit. Chapter 1 offers suggestions about how you might go about making that decision.

Your approach will be slightly different when you are commissioned to renovate an existing building. In that case, extensive examination of existing construction documents and field conditions is called for. If weather barrier failures have contributed significantly to the reasons for the renovations, you might want to consider seeking professional assistance with them throughout the design process. In that event, unless you have extensive experience with the subject, read Chapter 1 first. Then refer to other chapters as needed during the design and document production process. Even when you have a consultant on board, it is useful to know what to expect that consultant to do.

Building Contractors

How you use this book depends on which hat you are wearing at the time. For your own buildings, the suggestions given above for owners apply, except that you will probably have more experience with such problems than many owners do.

When you are asked by a building owner to repair a failed weather barrier, your approach might be very similar to that described above for architects and engineers.

When repairs to a weather barrier are part of a project for which you are general contractor, your problem becomes one of supervision of the subcontractor who will actually repair the failed barrier. It is always helpful to be able to refer to an authoritative source, such as those listed in this book, to double-check methods and materials your subcontractor proposes. It is also useful to be able to verify a misgiving you might have about the specified materials or methods. In each of those cases, read the chapter covering the subject you are interested in and check the other resources listed whenever a question arises. When deciding which subcontractor to use, review Chapter 1.

Even if you have extensive experience in repairing failed weather barriers, unusual conditions frequently arise. When they do, turn to the list of sources at the end of each chapter and the Bibliography to discover who to seek advice from.

Disclaimer

The information in this book was derived from published data of, and statements made to the author by representatives of, trade associations, standards setting organizations, manufacturers, government organizations, consultants, and architects. The author and publisher have exercised their best judgment in selecting data to be presented, have reported the recommendations of the sources consulted in good faith, and have made every reasonable effort to make the data presented accurate and authoritative. But neither the author nor the publisher warrant the accuracy or completeness of the data nor assume liability for its fitness for any particular purpose. Users bear the responsibility to apply their professional knowledge and experience to the use of data contained in this book, to consult the original sources of the data, to obtain additional information as needed, and to seek expert advice when appropriate.

Acknowledgments

A book of this kind requires the help of many people to make it valid and complete. I would like to acknowledge the many manufacturers, producers associations, standards setting bodies, and other organizations and individuals whose product literature, recommendations, studies, reports, and advice helped make this book more complete and accurate than it otherwise would have been.

At risk of offending the many others who helped, I would like to single out the following people who were particularly helpful:

Bob Biner, International Masonry Institute, Washington, D.C.

John Brennan, Asphalt Roofing Manufacturer's Association, Rockville, Maryland

Anthony M. Caccavale, W. R. Grace and Company, Beltsville, Maryland

Dell Ewing, President, Dell Ewing Associates, Inc., Vienna, Virginia

Dave Faulkner, Boral Concrete Products, Inc. (Lifetile), Lake Wales, Florida

Dick Fricklas, The Roofing Industry Education Institute, Englewood, Colorado

Earle W. Garner, President, Every Water Guard Co., Inc., Rockville Maryland

Alan D. Grayson, Associate Executive Director, Education and Administration, National Roofing Contractors Association (NRCA), Rosemont, Illinois

William G. Harrill, Technical Representative, Hey'di American Corp., Virginia Beach, Virginia

Robert Hund, Managing Director, Marble Institute of America, Farmington, Minnesota

Edward W. Johnson, CAE, Executive Director, Sealant Engineering and Associated Lines (SEAL), San Diego, California

Ray Kogan, Sigal Construction Corporation, Washington, D.C.

Bruce McIntosh, Portland Cement Association, Skokie, Illinois

William C. McLellan, Regional Sales Manager, Lifetile, Boral Concrete Products, Inc., Lake Wales, Florida

Harald Muench, Vice President, Hey'di American Corp., Virginia Beach, Virginia

Robert Phillips, President, Roof Consultants Institute (RCI), Raleigh, North Carolina

Eric Rickert, Sigal Construction Corporation, Washington, D.C.

Gerry Rousseau, Rock of Ages Corporation, National Building Granite Quarries Association, Barre, Vermont

Everett G. Spurling, Jr., FAIA, FCSI, Bethesda, Maryland

John Van Wagoner, Vice President, Prospect Industries, Sterling, Virginia

And I would like to especially thank the following people for their special help:

Thomas P. Carney, Advertising Manager, The Garland Company, Inc., Cleveland, Ohio

John W. Harn, Harn Construction, Co., Laurel, Florida

Kimberley A. Hollenkamp, Hydrozo Marketing Corporation, Manassas, Virginia

Richard Scruggs, W. R. Grace and Company, Baltimore, Maryland

Sally Sims, Librarian, National Trust for Historic Preservation Library and University of Maryland Architectural Library, College Park, Maryland

CHAPTER

1

Introduction

To protect the building and the people, equipment, and furnishings housed in it, every building must exclude wind and water and be free from extreme temperature variations. Walls, roof structures, and devices such as windows and doors that close openings in the shell are the principal barriers to unwanted weather. But alone they are seldom effective. Many shell materials are not weatherproof. They must be coated or covered with weather barriers. Joints must be sealed. Water that intrudes in spite of all efforts to exclude it must be lead back to the exterior. Water vapor must be prevented from condensing on interior surfaces or, worse yet, within the construction or on structural elements. An effective weather barrier is, then, a system of barriers to wind, water, and heat flow. No one part of the system will succeed without the other parts.

Almost every building will, at some time, experience a weather barrier failure. Most of the failures will manifest themselves as water damage. When a weather barrier failure occurs, there is both an immediate problem and a long-range problem.

The first problem, of course, is to prevent as much damage as possible to the structure and its contents.

The second problem is to prevent the failure from recurring.

This chapter discusses the nature of weather barriers and outlines steps a building owner can take to solve both the short- and long-range problems associated with weather barrier failures. This chapter also discusses the relationship of architects, engineers, building contractors, and roofing and waterproofing consultants to the owner on projects involving weather barrier failures and outlines ways those professionals can approach the problem-solving process in an orderly way.

This chapter includes an approach to determining the nature and extent of weather barrier failures and suggests the type of help a building owner might seek to help solve the problem.

Chapters 2 through 8 contain detailed remedies and sources of additional data. This chapter will help a reader determine which chapter to turn to next.

What Weather Barriers Are and What They Do

The term "weather barriers" in this book refers to condensation controls, water infiltration barriers, and heat flow barriers, which are not part of the structure or shell construction of a building. All must function properly to prevent weather barrier failure and damage to building structure, finishes, and contents.

The term "weather barrier failure" in this book refers to a failure of one or several of the weather barriers discussed.

Water Infiltration Barriers

Water infiltration barriers prevent water intrusion into buildings from the exterior. Those discussed in this book are roofing, membrane waterproofing, clear water repellents, dampproofing, and flashings.

This book does not discuss such weather barriers as building shell construction; masonry, stone, or concrete walls; siding; sheathing; glazed metal or wood wall or roof systems; doors, windows, or other rigid closers; or caulking and sealants.

Chapters 3, 4, 6, 7, and 8 contain detailed descriptions of the water infiltration barriers discussed in this book. The following general definitions apply.

Waterproofing: A membrane or homogenous material used to prevent water from reaching a protected surface. A homogenous material used to seal existing water leaks in building surfaces. Exterior applied waterproofing is usually concealed below finish grade or beneath paving or other finished surfaces. Interior waterproofing generally is used in locations where appearance is unimportant, such as parking garages and unoccupied basements. Where appearance is critical, interior applied waterproofing is usually concealed by finishes.

Clear Water Repellent: Painted or sprayed-on liquid compounds designed to help a building surface shed water before it has a chance to penetrate the surface. Water repellents are used to coat exterior surfaces. Decorative water repellents, which are usually colored, are beyond the scope of this book.

Dampproofing: A liquid or pasty bituminous material used to prevent dampness from reaching a wall surface or to prevent dampness which has penetrated a material from exuding into finished spaces. Dampproofing is used on exterior surfaces below grade and on interior surfaces behind finish materials. Dampproofing is not appropriate for surfaces subjected to hydrostatic pressure.

Roofing: A material that covers the rigid structural roof of a building and prevents water intrusion into the building.

Flashing: A sheet or membrane used to lead water to the exterior of a building. Flashing is used between roof and wall surfaces; around penetrations through waterproofing and roofing; at hips, ridges, and valleys in steep roofs; within walls above doors, windows, and other wall penetrations; and in other locations where water might enter the building through a juncture between two materials. Flashing is also used within walls to conduct intruded water back to the exterior.

Condensation Control

When air containing water vapor encounters a low enough temperature, the water vapor will change state (condense) into liquid water. If a surface is present at that point, the liquid water will collect on it in beads—a process which is called condensation.

Water vapor, present everywhere in varying amounts, migrates from zones of high vapor pressure into zones of lower pressure. In most of this country, at the times when condensation is most likely to be a problem (winter), the humidity inside buildings is higher than it is outdoors. That higher indoor humidity, which is mostly generated by people and their activities, will try to escape to the exterior. When it encounters a surface or space with a low enough temperature, condensation will occur.

Condensation can be prevented from forming on surfaces in occupied spaces by keeping the surfaces warm or by dehumidifying the air.

Preventing condensation in unheated attics and crawl spaces, and within wall and roof materials requires using insulation, vapor retarders, and ventilation by outside air to prevent conditions from occurring that will cause condensation to form.

Where condensation cannot be prevented, it is necessary to use wipe-off and weep systems which collect the water and lead it harmlessly to the exterior. Such wipe-off or weep systems are common in glazed curtain walls, windows, and both metal-framed and plastic skylights.

Chapter 2 contains a more detailed discussion of condensation problems and their solutions.

Heat Flow Barriers

Heat flow barriers reduce heat flow in or out of buildings. The most-used heat flow barriers are thermal insulation products manufactured specifically for that purpose. Properly contained air spaces are also effective thermal barriers. Insulating glass, for example, relies on contained air space for much of its thermal transmission resistance. Dense materials, such as solid masonry and poured concrete, are also heat flow barriers, but they are not as effective as are materials that contain air spaces.

The thermal insulation discussed in this book is that used in roofing systems. Insulation used between exterior wall studs; behind finishes on the interior face of exterior masonry, stone, and concrete; in attic spaces; within concrete masonry units; between the wythes of masonry cavity walls; in cavities between interior finishes and exterior surfaces; in cold storage facilities; above ceilings and in partitions for sound reduction; as part of exterior insulation and finish systems; or for other purposes are beyond the scope of this book.

Failure Types and Conditions

The most obvious sign of a failed weather barrier is water dripping from a ceiling. Walls painted green with mildew are easy to see, too (see Fig. 1-1). But evidence of many weather barrier failures is more subtle, perhaps just a touch of efflorescence or a spalled concrete window sill. Recognizing some weather barrier failures requires no special expertise. Assessing the cause of the failure and determining the proper remedy, however, often requires a detailed knowledge of weather barriers.

Most weather barrier failures fall into one of four categories: condensation, water infiltration, wind damage, and thermal barrier inconsistencies.

Water Infiltration

Visible water is one evidence of water penetration through a water infiltration barrier but not necessarily the most common one. More commonly, water

Figure 1-1 A gutter leak resulted in this mildew on a south Florida wall. (*Photo by author.*)

leaks manifest themselves as stains such as those caused by rust, color changes, fungal growth, such as mildew, spalling or surface delamination, splitting or crazing of finished surfaces, and efflorescence (see Fig. 1-2). Visible water is as likely to be a condensation problem as a water infiltration barrier failure.

Most apparent water damage, and almost every instance of visible free water that is not condensation, results from leaks that occur above the site of the observed water. But some water penetration can come from below through capillary action, and some can come directly horizontally through wicking action or wind-driven rain passing around a water barrier. It is important to determine the direction of the penetration, because that knowledge will help determine the source of the leak.

Leaks do not always occur at the site of the visible damage. Water can migrate through the structure for many feet before becoming apparent. Thus a parapet mortar joint failure can mimic a flashing leak. It is important to

Figure 1-2 Efflorescence—a sign of a failed weather barrier. (*Photo by author.*)

trace the leak to its source. The task is sometimes difficult and may require professional knowledge.

Water damage is not always apparent immediately after the leak occurs. Minor leaks can cause major damage over time. Mildew does not appear immediately. Efflorescence may not occur until after a heavy rainfall, even though the underlying construction is damp for an extended time.

It is important to determine whether a leak is caused by water infiltration barrier failure or failure in other enclosing elements. Leaks often occur through masonry cracks and faulty mortar joints, especially in parapets, chimneys, and other masonry projecting above an adjacent roof. Other major sources of leaks are prefabricated or preassembled building elements such as glazed curtain walls and skylight-type structures. Preformed roofing and siding sometimes leaks. Leaks occur in doors, windows, louvers, hatches, vents, and other items mounted in walls or roofs. Leaks between materials or between items penetrating the building enclosure and the enclosure components (metal roofing, for example) are sometimes flashing or sealant failure. Flashing failures are discussed in Chapter 8.

It may not be immediately apparent which part of the weather barrier system has failed when leaks become visible (see Fig. 1-3). Because water tends to migrate to low points, leaks through defective gutters, flashings, or sealant joints may appear to be roof leaks. It is necessary to trace the leak to its source.

Condensation

Condensation is a water infiltration problem when it occurs in a crawl space or attic, but the remedy is usually more like one used for a condensation problem than an infiltration problem. The solution usually lies in a combination of a vapor barrier and ventilation.

Condensation inside a building, inside wall or roofing materials, or on structural elements is a weather barrier failure. It may be an insulation problem, an underlying construction failure, a ventilation problem, or a vapor barrier problem. Under some conditions, condensation is so profuse

Figure 1-3 This "roof" leak was actually caused by a failed gravel stop. (*Photo by author.*)

that it mimics major exterior water penetration. Where humidity is high, such as in shower rooms or kitchens, or in attic and crawl spaces in very hot and humid climates, condensation can occur on fairly warm surfaces. Condensation on interior surfaces of exterior wall components or within materials used as exterior wall components suggests insufficient insulating qualities through the wall and a lack of proper vapor barriers.

Condensation is most harmful when it occurs within a material. That happens when the insulation value is such that the dew point occurs within a material that is on the warm side of a vapor barrier or where there is no vapor barrier.

Wind Damage

Wind damage is usually obvious to anyone who can see the affected surface (see Fig. 1-4). Sometimes wind damage occurs in locations a building owner seldom visits, such as on low-slope roofs. Periodic inspections of areas not usually visible will often reveal wind damage before the damage is serious enough to cause leaks.

Figure 1-4 The worst kind of wind damage to a roof. (*Photo courtesy of The Garland Co., Inc.*)

Wind damage often occurs where the damaged material was weakened by other causes before the wind damage. Old asphalt shingles are much more likely to suffer wind damage than are new shingles, for example.

Thermal Barrier Inconsistency

Thermal barrier inconsistencies are instances where the thermal barrier is not fully doing the job it was installed to do. Such failures are due to one of several causes.

Probably the most-encountered form of thermal barrier failure is open passages that permit heat to flow directly from interior to exterior, or vice versa. Failed joint sealants and weather stripping are common reasons. Door and window perimeter caulking and weather stripping are prime culprits.

Probably the second most common thermal barrier failure is due to loss of roof or wall insulation's thermal value from water intrusion into the thermal insulation.

Thermal barrier failures also result from initial improper installation of an insulating material. Either the surface was not fully covered with insulation or the insulation was improperly supported and has sagged, leaving openings in the insulation envelope.

Thermal barrier failures are not often easy to detect, sometimes requiring special devices to measure heat flow or removing portions of the covering material to be able to observe the insulation directly.

The Next Step

When the type of weather barrier failure has been determined and the failed barrier has been tentatively identified, turn to the appropriate chapter of this book or the other book in this series which addresses the identified material. After reading the material there, proceed to solve the problem in the appropriate way. If after reading the material there, you realize that you are not competent to proceed further, seek professional help.

What to Do in an Emergency

In an emergency, don't analyze. Act.

When action must be taken immediately to stop damage that is already occurring or imminent, do whatever is necessary. Action varies from placing buckets to catch the water to covering the failed barrier, when it can be immediately identified, with an impervious material. Unfortunately, most of the time it is not practicable to fix leaks in weather barriers until the weather conditions that caused the immediate problem have abated.

A word of caution, however. Do not, unless absolutely unavoidable, take an irreversible remedial action. Small repairs that cannot be easily removed can become major cost items when permanent repairs are attempted. Coating a leaking roof with an incompatible sealer may not only fail to solve the immediate problem, but may require removal of the coated roofing entirely when the rain stops.

If water is pouring through a hole in a basement wall, stuff something in the hole. Worry about where the water originates after the water stops coming in.

Professional Help

If after reading the chapter of this book or the other book in this series that addresses the identified problem area, the next step is still unclear, or it is not possible to be sure of the nature of the problem, consult with an expert in weather barrier failures. Who that is may vary with the knowledge and experience of the person seeking the help.

Help for Building Owners

Building owners can turn to several levels of expert help with weather barrier problems.

Emergencies.   In emergencies, seek the most available professional help first. If the owner has a relationship with a building professional, that's the place to start. If the owner has relationships with several building professionals, the place to start is with the most appropriate. For example, suppose the leaking roofing was recently applied over an existing building by a roofer under a direct contract with the owner. That is, there was no general contractor involved. If the first time it rains, the roof leaks, obviously that owner should call the roofing installer.

The right person to call is someone who can remedy the immediate problem. Worry about protocol later. An owner with a leaking roof should go first to a roofer. The exception, of course, is if the building is still under warranty by a general contractor. Then the owner should go to the person who is responsible for the warranty, which is usually the general contractor.

Probably the last person to call in an emergency is an architect or engineer, unless the architect or engineer is the only building professional the owner knows. An architect or engineer might be able to identify an organization that could stop the damage from occurring. It is always better to seek help from a known organization than to simply thumb through the telephone book.

In some areas of the country, a call to a local organization representing building industry professionals will net a list of acceptable organizations specializing in repairing the kind of damage that is occurring. Such local organizations might include a local chapter of the Association of General Contractors (AGC), an association of roofing and waterproofing contractors, or the American Institute of Architects (AIA).

It might be possible to reach a local roofing or waterproofing consultant, but that person may only be able to recommend someone who can repair the damage.

In each case, the methods recommended later in this chapter for contacting the various entities would apply in an emergency. There would not be time, however, to do the double checking suggested.

Architects and Engineers.   Even in nonemergency situations, building owners with easily identifiable weather barrier problems do not usually need to consult an architect or engineer. An exception occurs when repairing a weather barrier failure that is part of a general remodeling, renovation, or restoration project. Even when the repairs are solely related to weather barrier failure, repairs that necessitate manipulation of structural systems or those which might cause harm to the public may require a building permit. In many jurisdictions, building officials will not accept permit applications unless accompanied by drawings and specifications sealed by an architect or engineer. Sometimes, the scope of the problem warrants hiring an architect or engineer to prepare needed drawings and specifications, even when authorities having jurisdiction do not require such documents.

When the owner hires an architect or engineer, the same standards one would usually follow when hiring such professionals apply. In addition to the usual professional qualifications applicable to any project, however, architects and engineers commissioned to perform professional services related to an existing building should have experience in the type of work needed. It is seldom a good idea to hire an architectural firm with experience solely in office building construction, no matter what the scope, to renovate an existing hospital.

General Building Contractors.   Simple weather barrier failure problems are often best handled by a building contractor. Depending on the scope, the contractor may be a general contractor or a specialty subcontractor. If the problem is isolated and involves a single discipline, a single specialty subcontractor may be all that is needed. If multiple disciplines are involved, a general contractor may be needed to coordinate the activities of several subcontractors. Owners with experience in administrating contracts may be able to coordinate the work without a general contractor.

When selecting a general contractor, it is best to select one who is known to the owner. If there is no such contractor, seek advice in finding a competent firm. Such advice might come from satisfied building owners, architects or engineers the owner knows, or the American Institute of Architects. As a last resort, an owner might ask for a recommendation from a contractor's organization. Bear in mind, however, that if you ask for a recommendation from an organization that is supported by and represents the firm being sought, you will, at best, get a list of firms in the area, which is not a good way to obtain a recommendation. This is in no way an indictment of contractors or their associations. The same advice applies to doctors, lawyers, and architects. You find the best ones by word of mouth.

If the project is large enough to warrant doing so, an owner might consider seeking competitive bids from a list of contractors. Few weather barrier repair projects are that large or complicated, however. Negotiation is better if reputable firms can be found to negotiate with. If all the parties are unknown, competitive bidding may be the only way to get a reasonable price.

Specialty Subcontractors. If the weather barrier repair project is small and not complex, a building owner is usually better off hiring a specialty subcontractor directly. Use the same methods suggested earlier for finding a reputable general contractor. In addition, general contractors the owner knows to be reputable are sources for recommendations for subcontractors, as long as there is no symbiotic relationship between the general contractor and the subcontractor.

Competitive bidding is a more acceptable method when dealing with a subcontractor for small projects than is the same arrangement with a general contractor.

Roofing and Waterproofing Consultants. Owners sometimes engage roofing and waterproofing consultants to determine the nature of a weather barrier problem, find a solution, select repair products, write specifications and produce drawings related to the solution, oversee the repairs, and, sometimes, even make the repairs themselves.

Unfortunately, all roofing and waterproofing consultants are not created equal. Many are building product manufacturers' representatives trying to increase their sales. While most of them are reputable, not all are competent to give advice on a broad range of solutions to weather barrier failure problems. Following the recommendations of an incompetent consultant can cause problems that will linger for years.

Selecting a roofing or waterproofing consultant can be an arduous project, filled with uncertainty and potential harm. Since there are no licensing requirements, anyone who chooses to do so can hang up a shingle that says "Roofing and Waterproofing Consultant." Fortunately, there are two associations representing consultants: the Roofing Consultants Institute (RCI) and the Institute of Roofing and Waterproofing Consultants (IRWC). In 1987, RCI began a certification program for roofing consultants which entails a 400-question written examination, education requirements, and experience requirements. The education and experience requirements are rated on a point system, which, within limits, permits a strength in one to offset a weakness in the other. The RCI examination is reported to be excellent and their qualifications requirements are sufficiently rigorous to weed out all but the best, but at the time this book was written, the program was very new. By mid-1988, RCI had certified only eight consultants.

A building owner lucky enough to be able to hire a RCI-certified consultant should, by all means, do so. Other owners will have to qualify the consultant themselves. One way to do that is to hire an architect or engineer and let that person select and qualify the consultant, subject to the owner's approval, of course. Asking a roofing or waterproofing contractor to recommend a roofing or waterproofing consultant may not be a good idea. Many roofing and waterproofing contractors feel that consultants are at best a necessary evil. Their opinion probably stems from the tendency of some manufacturers' representatives to oversell their abilities and knowledge.

Whoever does the hiring, roofing and waterproofing consultants should have the following qualifications:

- Membership in a nationally recognized organization representing water-proofing and roofing consultants. An owner might want to ignore this qualification if the potential consultant is a licensed architect or engineer.
- An extensive knowledge of roofing and waterproofing and experience with many different types of roofing and waterproofing. Of course, a roofing consultant need not have an extensive knowledge of waterproofing, nor does a waterproofing consultant need extensive roofing knowledge. A knowledge of tile roofing is not too critical if the building in question has only a built-up bituminous roof.

A good solution to finding a qualified roofing and waterproofing consultant is to select one recognized in your geographic area as an expert.

If the local grapevine fails, contact RCI or IRWC and ask for a list of their members in the area where the project is located.

Help for Architects and Engineers

Architects and engineers usually get involved in weather barrier repairs only when the repairs are extensive, have occurred on a prestigious building, or are part of a larger renovation, restoration, or remodeling project.

Architects who do not have extensive experience in dealing with weather barrier failures or working with existing weather barriers should seek outside help from someone who has such experience. The nature of that help and the person selected to consult depends on the type and complexity of the problem.

Other Architects and Engineers.   One source of professional consultation for architects and engineers is another architect or engineer who has had experience with the type of weather barrier problem at hand. The qualifications needed are similar to those outlined earlier in this chapter under the heading "Roofing and Waterproofing Consultants." The other architect or engineer need not be in architectural or engineering practice. Do not overlook specification's consultants or qualified architects and engineers employed by government, institutional, or private corporate organizations.

Roofing and Waterproofing Consultants.   Architects can also hire roofing and waterproofing consultants who are not architects or engineers when the situation warrants doing so. The qualifications outlined earlier in this chapter under the heading "Roofing and Waterproofing Consultants" apply no matter who is hiring the consultant.

Product Manufacturers.   Product manufacturers and their associations often provide sufficient information for a knowledgeable architect or engineer to deal with many weather barrier failures. The architect or engineer must, of course, compare product manufacturers' statements with those of other manufacturers and industry standards and exercise good judgment in deciding which claims to believe. The problem is the same as any other where architects and engineers consult producers and their associations for advice. The architect or engineer must study every claim carefully, especially if the claim seems extravagant, and double check everything.

Help for General Building Contractors

Whether a general building contractor needs to consult a waterproofing or roofing expert depends on the complexity of the problem, the contractor's own experience, and whether the owner has engaged an architect or roofing or waterproofing consultant. Duplication of effort is unnecessary unless the contractor intends to challenge the views expressed by the owner's consultants or the content of the drawings and specifications.

A general building contractor acting alone on a project where the owner has not engaged an architect, engineer, or consultant must base the need for hiring consultants on such factors as the contractor's experience and expertise with the types of problems to be encountered, the contractor's specialty subcontractor's experience and expertise in dealing with the types of problems involved, and the complexity of the problems.

Hiring a roofing or waterproofing consultant may complicate a general contractor's relationship with a roofing or waterproofing subcontractor. Such duplicity is seldom justified and usually is a bad idea. An experienced and qualified roofing subcontractor is not likely to appreciate a roofing consultant hired to tell the roofing contractor how to do the job. The general contractor would be better off finding a qualified subcontractor to do the work and rely on that subcontractor's advice. If no such subcontractor is available and the general contractor is not experienced with the problem at hand, hiring a consultant may be necessary, regardless of the feelings of the subcontractor. Muddling along to salve feelings is wholly incompetent and unprofessional. In those circumstances, the contractor should recommend that the owner employ a qualified architect or engineer to specify the repairs and let the owner and the design professional hire a roofing or waterproofing consultant if necessary.

In the unlikely event that a contractor should hire a roofing or water-proofing consultant, the earlier recommendations in this chapter apply.

Prework On-Site Examination

On-site examinations before the work begins are important tools in helping to determine the type and extent of a weather barrier failure and the resultant damage to underlying construction. Who should be present during an on-site examination is dictated by the stage at which the examination will take place.

The Owner

The first examination should be by the owner or the owner's personnel to determine the general extent of the problem. This examination should help the owner decide what the next step should be and the type of consultant the owner needs to contact, if any.

When the owner has selected a consultant, the owner and the consultant should visit the site and define the work to be done. The owner's consultant may be an architect or engineer, general contractor, specialty subcontractor, or roofing or waterproofing consultant. This second site examination should be attended by a representative of each expert the owner has engaged to help with the problem. A roofing or waterproofing consultant, if engaged by another of the owner's consultants, should also be present. The general contractor's specialty subcontractors should also be present, if they have been selected.

During the second site visit, the parties should become familiar with conditions at the site and offer suggestions about how to solve the problem.

Architects and Engineers

An architect or engineer hired to repair a failed weather barrier should, before visiting the site if possible, determine the products and systems used in the failed weather barrier. The architect should then visit the site with the owner to determine the extent of the work and to begin to decide how to solve the problem. Based on discussion with the owner about the nature of the problem, the architect should have decided whether to engage professional help. If a consultant is to be used, that consultant should visit the site with the owner and the architect.

If as a result of the architect's first site visit, the architect determines that a waterproofing or roofing consultant previously considered unnecessary is needed, the architect should arrange for another site visit with that consultant.

During the progress of the work, the architect and the architect's consultant should visit the site as often as is necessary to fully determine the nature and extent of the problem and to help arrive at a total solution. These site visits should extend the observation beyond the immediate problem to ascertain whether additional unseen damage might be present.

Building Contractors

Non-bid Projects.   On non-bid projects a building contractor may wear at least two hats.

The easiest situation to deal with is a negotiated bid based on professionally prepared construction documents. In that case, the contractor should conduct an extensive site examination to verify the conditions shown and the extent and type of work called for in the construction documents. Offering a proposal based on unverified construction documents is a bad business practice that can cost much more than proper investigation would have, if the documents are later found to be erroneous.

When the owner has not hired an architect or consultant to ascertain and document the type and extent of the work, the contractor must act as both designer and contractor. Then, the contractor should visit the site with the owner as soon as possible, and revisit as often as necessary, to determine the nature of the problem and the extent of the work to be done. A carefully drawn proposal is an absolute must to be sure that the owner does not expect more than the contractor proposes to do.

Even when the owner hires an architect or other consultant, the contractor should visit the site with the owner and the owner's consultant as soon as permitted. The purpose of the visit is to ascertain the extent and type of work to be done and to recommend repair methods. Invite specialty sub-contractors to also visit the site with the owner, the owner's consultants, and the contractor, if possible. The more input the contractor has in the design process, the better the result is likely to be.

Bid Projects.   Even when the contractor is invited to bid on a project for which construction documents have been prepared, a prebid site visit is imperative. No contractor should bid on work related to existing construction without extensive examination of the existing construction. Some construction contracts demand it. Some construction contracts even try to make failure to discover a problem the contractor's responsibility. Even if the courts throw that clause out, who can afford the time and costs of a lawsuit. A contractor should know the project well and establish exactly what work is to be done before bidding. Insufficient data may be cause for not choosing to bid on a project.

2

Condensation Control

Unless steps are taken to prevent it, at some time during each year, condensation will form on interior surfaces or within roofs or walls of most buildings, causing serious damage. Under some temperature and humidity conditions, condensation can be so heavy that it is mistaken for leaks from the exterior. This chapter discusses condensation problems that affect buildings and offers some suggestions for dealing with those problems.

What Condensation Is and Why It Occurs in Buildings

The following definitions apply when terms listed are used in this book.

Absolute Humidity: The actual weight of water vapor in air.

Condensation: Conversion of water vapor into water droplets; the water droplets so converted.

Dew Point: The temperature at which the air at a given pressure cannot hold any more water vapor (100 percent relative humidity) and

condensation begins. Air at the dew point is said to be "saturated."
The dew point is affected by pressure and humidity. The higher the
humidity, the higher the dew point temperature. Conversely, the
higher ~~lower~~ the air temperature, the more water vapor the air can contain
without being saturated.

Permeance: The property defining the rate at which water vapor migrates through materials. Permeance is measured in perms. According to ASTM Standard E96, one perm is the passage of one grain of
water vapor through one square foot of material in one hour for
each inch of mercury pressure differential between the two sides of
the material.

Relative Humidity: The ratio of the actual weight of water vapor in a
given volume of air to the maximum weight of water vapor that
amount of air can hold at the same temperature and pressure. A 50
percent relative humidity means that the air contains half as much
water vapor as it is capable of holding without condensation
forming.

Vapor Retarder: A material with a low permeance. "Vapor barrier," a
term used in the construction industry for many years, has been replaced by "vapor retarder," because no material is a true water vapor transmission barrier. The implication of the term vapor barrier
has often lead to misconceptions and errors.

Most sources agree that vapor retarders must have a perm rating
of 0.5 or better. The American Society of Heating, Refrigerating and
Air Conditioning Engineers 1985 *ASHRAE Handbook; 1985 Fundamentals*
suggests that a vapor retarder with a factory perm rating of 0.01 may
perform at no better than a 0.05 perm rating even when the joints are
carefully sealed. The *ASHRAE Handbook* also recommends a vapor
retarder perm rating of 0.05 for heavily insulated wood-framed roofs
without attics, which is more stringent than are some other guidelines.

Vapor retarders, especially in roofs, are not very good barriers to
water vapor migration. Even when the materials used have low perm
ratings, joints are difficult to seal and tears in the membrane are frequent.
Vapor usually migrates slowly enough so that vapor migration through
vapor retarders is not much of a problem, unless there are open joints
and unsealed openings in the retarder membrane. For vapor retarders
to be successful, the entire membrane must be continuous without open
joints or breaks. Joints around penetrating elements, such as pipes,
must be sealed.

Condensation will form anywhere the temperature dips below the dew
point. Water vapor migrates through almost every material from zones of

high vapor pressure to zones of low pressure. Since the migration occurs regardless of airflow, air leaks are not necessary for the vapor to permeate materials. In cold northern climates, the winter exterior vapor pressure is usually less than the interior pressure in a building. Water vapor will then tend to migrate from the interior to the exterior. In humid southern climates, the pressure outdoors may be greater than that in air-conditioned interior spaces and the vapor may tend to migrate from the exterior to the interior.

Thus, condensation can form anywhere the necessary combination of temperature and relative humidity occurs. The most commonly recognized manifestation of condensation is free water beading on or coating a surface such as a window or the walls of a shower room. But the most destructive condensation is that which occurs within materials and only becomes visible after those materials have been damaged.

Sources of water vapor in buildings include:

- Construction operations, such as drying of concrete, masonry, and other wet materials.
- Humans (or other animals), which generate water vapor by evaporation; building occupants also introduce copious amounts of water into the air from cooking, bathing—particularly when using showers—and other activities.
- Industrial activities such as manufacturing and commercial activities, such as food preparation, which generate high relative humidity levels inside buildings.
- Swimming pools, which evaporate large amounts of water into the air.
- Ground moisture, which can migrate upward into buildings through crawl spaces and through floors and walls.
- Humidifiers, which are designed to increase humidity.
- Loose-laid EPDM rubber roofing membranes, which are highly permeable when hot, permitting vapor entry when vapor pressure is lower inside than outside. EPDM membranes are less permeable when cold and can therefore trap some moisture in the roof system. However, there is disagreement among roofing professionals about whether the level of moisture introduced due to this phenomenon is sufficient to cause significant problems.

Condensation Control

Condensation can be a problem on exposed surfaces and in concealed locations as well. Different portions of buildings require different control methods, although the purpose of all methods is to prevent saturated air from reaching surfaces which are cold enough for condensation to occur.

Common Control Methods

High Interior Humidity Areas. In buildings, and portions of buildings such as commercial kitchens, spaces housing swimming pools, humidity-generating industrial facilities, and other locations where activities induce high levels of humidity, condensation on exposed surfaces is often reduced by mechanical ventilation and dehumidification. In such locations, vapor retarders are often placed around the building envelope, including walls and roof, and, sometimes, the floor.

Occupant Comfort. Dehumidification and mechanical ventilation is sometimes used to lower relative humidity to produce more comfortable spaces and prevent condensation on interior surfaces, even when very high relative humidity is not present. Air-conditioning, of course, lowers relative humidity.

In warm climates, dehumidification can actually contribute to condensation problems by lowering the interior vapor pressure even further, thus contributing to the rate of vapor migration from outdoors.

Exterior Walls. Condensation within exterior walls is eliminated by preventing the dew point from occurring on the warm side of a vapor retarder (see Fig. 2-1). Properly designed exterior walls are calculated so that the dew point for the expected humidity occurs within the insulation. A vapor retarder is introduced on the warm side of the insulation to prevent the high humidity from reaching the design dew point. Thus, humid air never reaches the dew point and condensation cannot occur.

To be effective, vapor retarders must be placed on the side of the insulation having the highest relative humidity. In cold climates for most of the year, that is on the interior side of the insulation. In hot humid climates, the vapor barrier might be more effective on the outside of the insulation, if the interior is air-conditioned. In both cases, calculations are necessary to determine the proper location of the vapor retarder.

Unheated Attic Spaces. In addition to other methods used, controlling condensation in unheated attic spaces requires ventilation. Codes usually dictate the amount of ventilation, but their requirements vary, may not cover all conditions, and in a few cases, may be inadequate. Ramsey/Sleeper, The AIA Committee on Architectural Graphic Standards' 1981 *Architectural Graphic Standards* recommends that the total net ventilation free area be 1/300th of the area of the attic where gable end ventilators can be used and an additional 1/600th at the ridge for conditions where gables do not exist, such as at hip roofs, or are inaccessible, such as in town houses. The *Architectural Graphic Standards* recommendations are generally in line

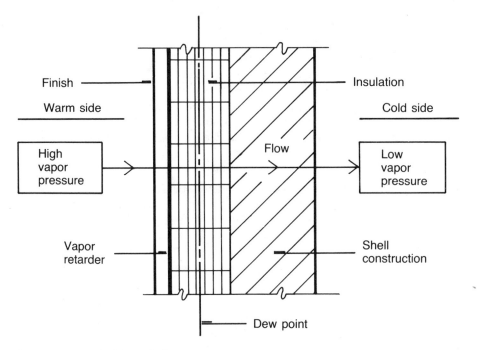

Figure 2-1 Exterior wall diagram.

with other industry recommendations. Some codes, however, reduce the required area when exhaust ports are at least 3 feet higher than intake ports.

In most of the United States, elimination of condensation in attic spaces requires placing a vapor retarder at the ceiling line. *Architectural Graphic Standards* recommends that vapor retarders be used everywhere. It is general practice, however, in warm regions to not install a ceiling vapor retarder beneath wood-framed roofs with slopes of 3 inches per foot or greater. In extremely humid climates, vapor retarders at ceilings beneath ventilated attics beneath high-pitched roofs may be contraindicated, because condensation may form on the outside of a ceiling vapor retarder whose temperature is below the dew point of the outside air.

If condensation in ventilated attic spaces is to be eliminated, moist indoor air must not be permitted to leak into the attic spaces, even when vapor retarders are not used at the ceiling line. Joints around pipes, ceiling hatches, and other items penetrating the ceilings must be completely sealed. The membrane must be continuous and complete.

Adding mechanical ventilation may increase interior comfort, but unless means are provided to ensure continuous ventilator operation, natural ventilation must still be provided.

Unheated Crawl Spaces.   Condensation in unheated crawl spaces is controlled by ventilating the space and installing vapor retarders.

Codes may dictate the amount of ventilation required for various crawl space conditions. *Architectural Graphic Standards* recommends that the total net ventilation free area in square feet be the sum of 1/100th of twice the length of the crawl space perimeter in feet plus 1/300th of the net crawl space area in square feet. Vents should be placed on each side of a crawl space and as high as possible.

In addition to ventilation, condensation prevention in crawl spaces requires placing a vapor retarder membrane on the ground surface and a second retarder membrane on the warm side of the floor insulation above the crawl space.

Roofs.   Besides roofs above unheated attic spaces, which are generally not insulated and need no vapor retarder, for purposes of this discussion, roofs fall into three categories: insulated framed systems, compact conventional systems, and protected membrane roof systems. Industry guidelines for condensation control are different for each category. In addition, the different guidelines do not all agree in their recommendations even within some of the categories. Usual condensation control methods for the three categories are as follows.

Insulated Framed Systems.   In insulated framed roof systems, an air space separates the roofing membrane from the insulation. Roofs in this category are more susceptible to condensation problems than are most other roof systems. Many such systems are wood-framed.

If it were possible to prevent moisture-laden air and water vapor from migrating from the building interior into insulated framed systems, ventilation would probably not be necessary. Vapor retarders are recommended at the ceiling line beneath such roofs in all climates (see Fig. 2-2). But it is almost impossible to prevent all air and vapor migration, even when a vapor retarder membrane is installed. Ventilation is, therefore, also needed.

Ventilation free area recommendations vary from 1/300th of the area of the space to 1/150th of the area of the space. Most guidelines agree that each joist space should be separately ventilated and that at least 1-1/2 inches of free air space should exist above the insulation. Some guidelines recommend connection of joist spaces using purlins perpendicular to the joists.

When insulated framed roofs are large, or other factors make good ventilation difficult, it may make sense to use an unventilated system with

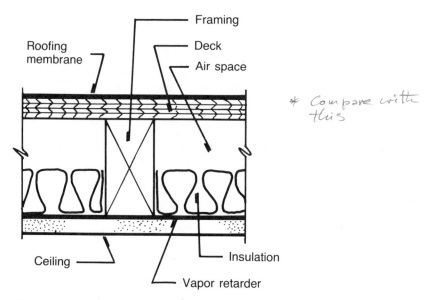

Compare with this

Figure 2-2   Insulated framed roofing system.

a vapor retarder and sufficient insulation above the roof deck to ensure that the dew point falls outside the enclosed space and outside the vapor retarder.

Compact Conventional Systems.   Insulated membrane systems where there is no separation between the insulation and the roofing membrane are called "compact" systems. Here, those compact systems with the membrane on the weather side are called conventional. Condensation in compact conventional systems is controlled by vapor retarders in the roofing system (see Fig. 2-3). The calculated dew point should fall on the low vapor pressure side of the vapor retarder. The circumstances under which insulated membrane roofing systems should include vapor retarders are controversial.

The National Roofing Contractors Association 1986 *NRCA Roofing and Waterproofing Manual* recommends the generally accepted criteria that vapor retarders should be used when the outside January average temperature is below 40 degrees Fahrenheit and the interior winter relative humidity is 45 percent or more. The NRCA recommendation has been in the past generally accepted in the roofing industry, but Wayne Tobiasson points out in his 1987 paper "Vents and Vapor Barriers for Roofs" that more definitive guidelines are available if one wants to investigate further.

Dick Fricklas, director of education for the Roofing Industry Educational

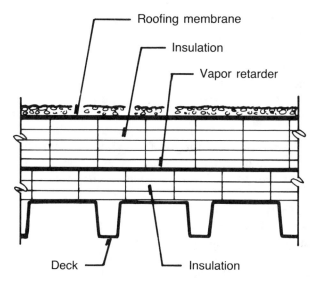

Figure 2-3 Compact conventional roofing system.

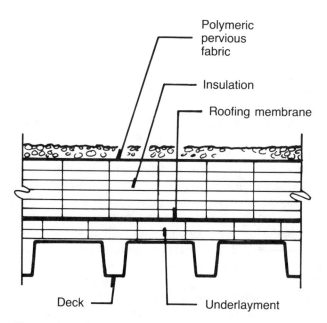

Figure 2-4 Protected membrane roofing system.

Institute, says in his 1988 editorial "New Guidelines for Specifying Vapor Retarders" that it is probably better to base vapor retarder use on vapor drive maps such as those included in Wayne Tobiasson and Marcus Harrington's publication "Vapor Drive Maps of the U.S.A." than on the conventional temperature-humidity criteria mentioned above.

The NRCA *Roofing and Waterproofing Manual* recommends that insulated membrane systems with vapor retarders be ventilated. Not all roofing experts agree, however. So, many roofs with vapor retarders remain unventilated. In "Vents and Vapor Barriers for Roofs," Wayne Tobiasson discusses the other point of view in some detail and gives references to support the lack of need for ventilation of insulated membrane systems.

Protected Membrane Roof (PMR) Systems.   PRM systems are compact systems in which the insulation is on the weather side of the membrane (see Fig. 2-4). In PMR systems, the roofing membrane, lying on the warm side of the insulation, acts as a vapor retarder. In cold climates, if the insulation is properly designed, there should be no condensation. Careful study is necessary, however, before a PMR is used in extreme southern climates over air-conditioned buildings where the indoor vapor pressure may be lower than the outdoor.

Condensation Control Failures and the Results

Condensation can do grievous harm to building materials, rotting wood, rusting metals, and destroying insulation efficiency. When the temperature in the space in which the condensation lies drops below freezing, ice crystals form. Ice expands and splits materials it has formed within, spalling concrete, cracking masonry, and delaminating roofing. In open spaces, ice can collect in large masses. When the ice melts in the spring, the water flows into surrounding construction, damaging everything in its path. A repeating freeze–thaw cycle can destroy materials as well. Glass block insulation is particularly susceptible, sometimes turning to powder after several cycles.

Condensation can contribute to decay of organic roofing materials. When condensation between roof insulation and the roofing membrane, between roofing plies, or within the lap seams of single-ply roofing evaporates, it can blister the membrane (see Fig. 2-5). Freezing condensation between built-up roofing plies can also delaminate the plies. Freezing condensation beneath adhered single-ply roofing or in the lap seams of a single-ply roofing causes loss of adhesion. Even condensation that forms on the bottom of the roofing package can drip onto and damage lower materials such as ceilings and light fixtures.

Weather barrier failures that manifest themselves as free water are easy

Figure 2-5   Roofing membrane blisters. (*Photo courtesy of The Garland Co., Inc.*)

to detect. It is hard to ignore water dripping from a ceiling. Some forms of water damage appear later as fungal growth, dry rot in wood, or rust on metal surfaces. Plaster may crack or crumble. The surface of gypsum board may swell and pull away from the core. Gypsum board panels may fall from ceilings. Paint may crack or peel. Concrete sills may spall. Masonry walls become coated with efflorescence. Mildew may paint walls green.

It is not always easy, however, to determine whether obvious water damage is the result of condensation or a leak. Both involve water intrusion into materials or spaces in the building. Water damage will appear about the same regardless of the cause. Severe leaks will, of course, be obvious. Smaller leaks may deposit no more water than does condensation.

In the interest of limiting damage, unless obviously condensation, it is best to assume that visible water is the result of a leak. Refer to the chapter of this book covering the material that is probably leaking. If the leak is not through one of the weather barriers discussed in this book, it will be necessary to examine other materials and systems. Leaks through masonry parapets and walls are common. Plumbing leaks and leaks through interior waterproofing, such as that below shower pans, should be eliminated as causes of visible water or water damage. Caulking and sealant joints should be carefully examined to ensure that leaks through them are not the source of the water.

When leaks have been eliminated as the source of the water, look for condensation control system failure. Most condensation control system failure stems from a fault in the original design or construction. Repair may require reconstruction. If, for example, a major thermal leak is discovered in an exterior wall and investigation reveals that the insulation in the wall construction is wet, the only solution is to remove the insulation and install new insulation. But that will not alleviate the problem. Unless the reason for the condensation forming is found and corrected, condensation will simply form again and saturate the new insulation.

Leaks are often easier to fix than are condensation problems. Solving condensation problems often requires major reconstruction. If, for example, condensation occurs because a vapor retarder was not installed between the finish material and the insulation in an exterior wall, there is no cheap solution.

A word of caution about the next part of this chapter. Taking anything other than the simplest actions to alleviate condensation problems without professional knowledge about such problems can be costly and ineffective. Virtually every time a major condensation problem occurs, especially when the problem is related to the building's roof, the sooner the owner seeks professional help, the less costly solving the problem is likely to be.

Solving Condensation Control Problems

Condensation failures can result from any one or a combination of the following.

High Indoor Humidity with an Inadequate Ventilation System. Solving problems resulting from high humidity and inadequate ventilation is easy, though not necessarily inexpensive. The statement of the problem implies that the condensation is on interior surfaces and not within wall, ceiling, or roof construction. The obvious solution is to improve the ventilation system. A simpler, and perhaps cheaper, method is to add a dehumidifying system. Both may be necessary. Before taking either step, calculate the results to see if the steps will solve the condensation problem. If they will not, then one of the steps outlined below may be necessary.

Not Enough Insulation in an Exterior Wall. "Not enough" is not subjective when one speaks of insulation. Independent of other considerations, which themselves may require more insulation, exterior wall insulation's "R" value should be such that the dew point does not fall on the high pressure side of the vapor retarder (see Fig. 2-1). If it does, and condensation is present, calculate the result of lowering the humidity inside the building and increasing ventilation within the affected areas. Dehumidifying the space and increasing circulation might solve the condensation problem, even though it will do nothing to alleviate the separate energy cost problems caused by too little insulation.

If dehumidification and increased ventilation will not prevent further condensation, it may be necessary to remove the finish material and vapor retarder, if any, on the warm side of the insulation and install additional insulation. Then add a new vapor retarder and finish. Adding insulation inside the old vapor retarder without removing the old vapor retarder could result in formation of condensation on the face of the old vapor retarder, even when a new vapor retarder is present. No vapor retarder is a perfect barrier.

Insulation that has become wet should be removed and replaced with new insulation. This could be a minor or major project, depending on the amount of insulation involved. The earlier the problem is detected, the less costly its solution will be.

No Vapor Retarder or Vapor Retarder on the Wrong Side of Wall Insulation.

When a vapor retarder is missing or incorrectly placed, first calculate the results of improving ventilation and adding dehumidification. In moist climates, the calculations will probably show that dehumidification and increased circulation will not produce the desired result. Then, the only solution will be to remove or significantly destroy the effectiveness of the old vapor retarder and install a new one in the proper location (on the warm side of the insulation). Where no vapor retarder is present, one must be added. Either will be a major project, entailing removal of existing finishes. Removing a misplaced vapor retarder may also require removing the existing insulation.

Wet insulation should be discarded and new insulation installed.

Moisture-Laden Air above Ceilings.

The American Society of Heating, Refrigerating and Air-Conditioning Engineers 1985 *ASHRAE Handbook; 1985 Fundamentals* instructs that moisture diffusion seldom causes condensation in roofing, that the more likely cause is moisture-laden air, and that the solution to condensation problems in insulated membrane roofing is to prevent air from bringing moisture to the roofing.

Preventing moisture-laden air from reaching the roofing requires sealing openings through which air may intrude into the space above ceilings. Openings occur not just through ceilings, and around pipes (see Fig. 2-6), ducts, conduits, light fixtures, ceiling hatches, and other penetrating items,

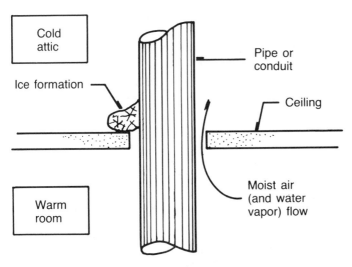

Figure 2-6 Unsealed penetration through ceiling.

but also from the interior of hollow construction penetrating the ceilings, including masonry, wood and metal stud, and other hollow walls and partitions (see Fig. 2-7).

Air leaks through floors over crawl spaces may also permit moisture-laden air to enter the crawl space where it will condense.

When air leaks are present, the materials around the joints may show signs of moisture damage. In winter, there may be water beads or ice around the opening.

The solution to air leaks, of course, is to seal every leak. Adding a vapor retarder membrane will sometimes help but may require removing existing finishes. Vapor retarders must be properly sealed also.

Leaking Vapor Retarder. Air leaking through vapor retarders is a major contributor to condensation in walls, crawl spaces, attics, and roofs. Leakage may result from damaged membrane, unsealed joints in the membrane, or unsealed membrane perimeter; or from failure to seal around penetrations such as pipes, ducts, electric outlets, conduits, light fixtures, ceiling hatches, floor hatches, ceiling access panels, and over the tops of hollow walls and partitions and at the ends of open partitions intersecting exterior walls.

The solution to a leaking vapor retarder is to seal the leaks. Doing so, however, may not be easy. It is difficult to determine exactly where leaks in a vapor barrier are. Obvious damage and ice formation, as discussed in "Moisture-Laden Air above Ceilings," may be fairly easy to spot, but holes, open seams, and unsealed perimeter conditions may be hard to find. Repairing

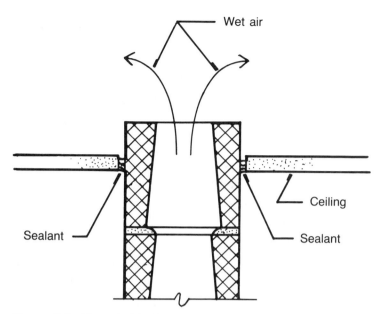

Figure 2-7 Unsealed hollow partition.

a vapor retarder at a floor, wall, or low-slope roof may necessitate removing the finish. In attics, it requires lifting walkways and insulation.

Not Enough Attic Space Ventilation. Of the several possible causes of condensation in attics, the easiest to check is ventilation. First examine the existing vents to see if they are blocked by insulation, animal nests, or something else. Remove blockages, if any. If that does not solve the problem, calculate the required ventilation area and verify that enough net square footage exists. If not, take steps to increase the area. Forced (power) ventilation only works while the power is on and is, therefore, a poor solution to condensation problems in attic spaces.

No Attic Space Vapor Retarder. When it is clear that lack of proper ventilation is not causing condensation in an attic, check to see if there is a vapor retarder at the ceiling level. If not, there are several possible next steps. One is to calculate the effect of lowering the humidity in the occupied space below. If the calculations show that to be a solution, install a de-humidifier. It may also help to increase circulation in the interior space.

If dehumidification and increased circulation do not solve the problem, and they probably won't, install a continuous vapor retarder on the warm side of the ceiling insulation and seal around all penetrations through the retarder membrane. Installation methods depend on ceiling construction.

If there is a vapor retarder in place, check to ensure that penetrations, joints, and perimeter are properly sealed, as discussed earlier.

Inadequate Crawl Space Ventilation. Regardless of the cause of condensation in a crawl space, proper ventilation is essential to reduce further incidents. Adequacy of the present ventilation should be routinely checked. Measure the net amount of existing ventilation opening and compare it with the recommended calculated amount. If there is insufficient ventilation, increase the opening sizes before taking more drastic measures.

If adding more ventilation does not solve the problem, or calculations show that there is already sufficient ventilation, ascertain whether the moisture is coming from above or below. This will not always be easy. Obvious conditions, such as wet ground or boundary walls that are wet at the bottom but dry at the top, suggesting that the major moisture is migrating upward, may not be present. It is, however, much more likely that the problem stems from excess ground moisture rather than water vapor migration from above.

Excess Groundwater in a Crawl Space. Standing water on the ground surface in a crawl space may indicate a serious problem, the solution of which requires drastic measures. The water may come from ground runoff; a wall leak; or a leak from pipes or through the floor from a shower, kitchen, or other wet installation above; or it may be bubbling up from an underground spring. Removing the source of the water is, of course, the first step, and may, without further action, alleviate the condensation problem. Removing standing water may require installing intercepting structures or a permanent drainage system. Anyone who is not an expert in such matters should not attempt to solve standing water problems without professional help. Refer to Chapter 1 for suggestions about finding professional help.

No Vapor Retarders in a Crawl Space. When standing water has been completely and permanently diverted away from the building, install a vapor retarder over the entire ground surface in the crawl space, unless the professional who helped with the water removal advises otherwise. A permanent drainage system alone may not completely solve the problem.

Even when visible surface water is not present, unless the moisture causing condensation in a crawl space is obviously coming from above, and even without direct evidence that the moisture is coming from below, the simplest and least costly first step in stopping crawl space condensation is to install a vapor retarder membrane over the ground surface. Make sure all joints, penetrations, and the periphery are completely sealed.

If a properly installed vapor retarder over the ground surface does not solve the problem, the moisture condensing in the crawl space is most likely coming from above. First, attempt to reduce the humidity in the space above using dehumidification, increased circulation, or both. If that does not work, or is not practicable, install a vapor retarder on the warm side of the floor insulation.

If there is already a vapor retarder on the warm side of the floor insulation, check to see that all openings, joints, and the perimeter are properly sealed, as discussed earlier.

Condensation in Roofing Systems. Condensation in roofing systems causes blisters or cracks in the roofing membrane or collects and drips onto other materials causing wet, sagging ceilings, peeling paint, rusted metal, rotted wood, or dripping water.

Insulated Framed Systems (see fig. 2-2). Where an air space separates the roofing membrane from the insulation, condensation is most likely the result of either an air leak from below into the ventilated air space, as discussed earlier, or inadequate ventilation in the space between the insulation and the deck.

In the second case, there may not be enough opening area into the space, connections between joist spaces may have been omitted, some joists spaces may not be ventilated, or the space may be too large to properly ventilate. The obvious solution is to increase the amount of ventilation. Rooftop ventilators may be necessary. Sometimes, introducing forced (mechanical) ventilation into the space may relieve the problem. Of course, to remain effective, forced ventilation must be kept constantly running. Wayne Tobiasson discusses this method in "Vents and Vapor Retarders for Roofs."

If increasing the ventilation is ineffective, as some experts insist will probably be the case, and sealing openings which permit warm air to flow from below into the ceiling space does not prevent condensation, more drastic measures may be required. One such possible measure is to close all vents into the roof space and convert the roof into an unvented system. This will result in a roof with insulation between the joists, a vapor retarder above the decking (sheathing), and insulation and roofing membrane above the vapor retarder. The amount of insulation between the joists will be limited depending on the humidity in the building. Designing such a system is a job for a knowledgeable professional.

Compact Conventional Systems (see fig. 2-3). Recent evidence seems to indicate that when properly constructed this is the roofing system least likely to experience condensation problems. Condensation in compact conventional roofing systems can, however, occur when such systems are improperly built.

Among errors that may lead to condensation are omitting a vapor retarder where one should have been used, failing to seal openings through the vapor retarder membrane, failing to seal joints in the vapor retarder membrane, and not providing needed ventilation. None of these problems is easy to solve. Their solution depends on the actual construction involved and the conditions encountered, including temperature and humidity conditions inside and outside the building.

When condensation occurs in a compact conventional roofing system, the most probable cause is moisture-laden air in the space between the ceiling and the roofing. Water vapor adjacent to the roof deck, migrating upward toward lower vapor pressure, will eventually contact the roofing membrane which will act as a vapor retarder. If the roofing membrane temperature drops below the dew point, condensation will form on the membrane. Such a temperature drop will not happen if the proper amount of insulation is placed above the membrane, however. So the scenario requires a design or installation failure.

If condensation should ever occur in a compact conventional roofing system, it may cause sufficient damage to require reroofing. If may freeze and break joints in or delaminate the roofing membrane. It may migrate, saturate the insulation, drastically reducing its insulating qualities, and rot or rust the roof deck. It may drip onto ceilings or run down inside walls, eventually destroying everything there if not stopped.

Reroofing may solve the immediate problem but will probably not prevent condensation reoccurrence, even when the new roofing is properly constructed with moisture retarder membrane and vents, unless moisture-laden air is prevented from infiltrating the ceiling space.

It might be possible to remove moisture-laden air from the ceiling space by ventilating or dehumidifying the space, but such is seldom practicable. Sealing the leaks through which the air enters, as discussed earlier, is a much better and more practical solution.

Protected Membrane Roof (PMR) Systems.   In PMR, because the membrane lies on the warm side of the insulation, the roofing membrane itself serves as a vapor retarder. Condensation can still occur below the membrane if the roof insulation is underdesigned. If condensation does occur, the advice given above for sealing and ventilating the space between the ceiling and roof applies. If that does not work, the insulation thickness must be increased.

Where to Get More Information

The American Society of Heating, Refrigerating and Air-Conditioning Engineers 1985 *ASHRAE Handbook; 1985 Fundamentals* offers guidance on vapor retarders and roof vents.

Wayne Tobiasson explains and supplements the *ASHRAE Handbook* material in his "Vents and Vapor Retarders for Roofs." Tobiasson's paper also lists additional sources of information about vents and vapor retarders and relates specific recommendations for roof ventilation. It also offers solutions to condensation problems and lists sources of additional recommendations for specific methods. Tobiasson's paper is available from the Building Thermal Envelope Coordinating Council. It was also reprinted in the November 1987 *The Construction Specifier.*

Wayne Tobiasson and Marcus Harrington's *Vapor Drive Maps of the U.S.A.* is a publication developed to help designers determine the need for roof vapor retarders. It is available from the Cold Regions Research and Engineering Laboratory.

Forrest Wilson's 1984 book *Building Materials Evaluation Handbook* has a chapter titled "Moisture Penetration and Damage" which also includes a general discussion of condensation and condensation control.

Ramsey/Sleeper, the AIA Committee on Architectural Graphic Standards 1981 edition of *Architectural Graphic Standards* discusses condensation and vapor retarders in some detail. Look up "Condensation of water vapor" in the index.

Architectural Graphics Standards also discusses roof and crawl space ventilation in detail.

The National Roofing Contractors Association's 1986 edition of *The NRCA Roofing and Waterproofing Manual* is an excellent source of data about condensation and vapor retardation related to roof construction. It includes general information and fundamentals about condensation, venting, and vapor retarders for roofs; diagrams and explanations to help a designer calculate the temperature at the vapor retarder level for various roof constructions; a psychrometric chart and other data necessary for calculation of dew-point temperatures; a discussion of various vapor retarder materials including recommendations and factors to consider when selecting one; suggested application methods and precautions for bituminous vapor retarders, laminated kraft paper vapor retarders, vapor retarders over high-humidity interiors, and vinyl-film vapor retarders.

The Brick Institute of America's *Technical Notes on Brick Construction,* "7C—Moisture Control in Brick and Tile Walls: Condensation," and "7D—Moisture Control in Brick and Tile Walls: Condensation Analysis" discuss condensation on walls and within masonry assemblies and include data useful in calculating where condensation is likely to occur.

Bituminous Dampproofing and Clear Water Repellents

This chapter discusses bituminous dampproofing and clear water repellents, suggests some guidance for their use, and offers recommendations for extending them, applying them over existing surfaces, and repairing them where appropriate. But before discussing clear water repellents and bituminous dampproofing in detail, it is necessary to show how they differ from waterproofing.

Waterproofing is a membrane, coating, or sealer used in concealed locations to prevent water from entering or passing through either horizontal or vertical building materials. Waterproofing is designed to exclude water even when the water is under hydrostatic head. Refer to Chapter 4 for further discussion.

Bituminous dampproofing is a bituminous coating used to prevent building materials from absorbing adjacent moisture or to prevent absorbed moisture from migrating farther into the invaded system. Dampproofing is not intended to prevent penetration of water under hydrostatic pressure. It is used on the exterior below grade and on interior surfaces. Both interior and exterior dampproofing, like waterproofing, are almost always concealed.

Clear (colorless) water repellent coatings are intended to reduce water penetration into building materials by capillary action. They are used on exterior wall surfaces above grade, or to prevent intrusion into and damage of horizontal concrete by water and sodium chloride or other ice-melting chemicals. Some clear water repellent coatings may prevent soiling and staining and are routinely used on limestone and concrete for that purpose. Some are advertised as aids in preventing graffiti from adhering, making marked surfaces easier to clean. They are sometimes used on architectural concrete to enhance the color. Clear water repellent coatings will not prevent passage of water under pressure.

Table 3-1 shows the types of surfaces on which waterproofing (WP), bituminous dampproofing (DP), and clear water repellent coatings (CWRC) are used.

Most bituminous dampproofing is applied on new materials. It is seldom applied to existing surfaces. When water penetrates below-grade walls, waterproofing is usually the remedial material choice. Existing bituminous dampproofing does sometimes need repair, but only rarely. Building additions, however, often require extending existing bituminous dampproofing.

Clear water repellent coatings are applied both on new surfaces and on existing wall surfaces that show evidence of water intrusion. Clear water repellents are often misused, causing more problems than they solve.

Paint and pigmented (opaque or semiopaque) coatings may function as water repellents or dampproofing, but are, nevertheless, intended primarily as decorative materials. Discussion of them is beyond the scope of this book.

Horizontal concrete coatings intended primarily to aid in the curing process are beyond the scope of this book, as are membranes intended to provide traffic-bearing surfaces or purely as decorative elements.

Table 3-1   Waterproofing, Dampproofing, and Clear Water Repellent Coatings Uses

	WP	DP	CWRC
Exterior above grade			X
Interior face of exterior wall		X	
Foundation wall			
Hydrostatic head	X		
No hydrostatic head		X	
Basement wall			
Hydrostatic head	X		
No hydrostatic head		X	
Elevator pit	X		
Tunnel	X		

Chemical dampproofing mixed into concrete is also beyond the scope of this book.

Bituminous Dampproofing Types and Their Uses

Bituminous dampproofing materials fall into one of two categories: hot-applied or cold-applied. Table 3-2 is a chart showing the conditions under which each type is usually used.

Hot-Applied Bituminous Dampproofing

Hot-applied bituminous dampproofing may be either coal-tar or asphalt based. Both are used in exterior below-grade applications. Coal-tar is usually preferred. Hot-applied asphalt dampproofing is occasionally used in interior applications. Coal-tar is not often used indoors because it has a distinctive unpleasant odor that, while it does diminish over time, never completely disappears. Coal-tar and asphalt bitumens are not compatible and should not be used in the same installation.

Hot-applied dampproofing is used below grade on exterior building walls where occupied space occurs on the interior side of the wall, and on foundation walls where some dampness is expected in the adjacent soil. Hot-applied bituminous dampproofing should never be used when water under hydrostatic pressure will occur. Hot-applied dampproofing is mostly used on large installation over smooth concrete or masonry coated with cement or mortar parging to create a smooth dense surface.

Both the Brick Institute of America's *Technical Notes on Brick Construction*, "7—Dampproofing and Waterproofing Masonry Walls" and the National Concrete Masonry Association's "NCMA—TEK 55" recommend that below-grade hot-applied dampproofing on masonry be applied over parging.

Concrete to receive hot-applied dampproofing will not need a coat of

Table 3-2   Hot-Applied versus Cold-Applied Bituminous Dampproofing

Criteria	Hot-Applied	Cold-Applied
Exterior		
Large Project	X	
Small Project		X
Interior		X
Where solvents are prohibited		X
Where breathing is required		X

parging unless the wall is honeycombed, pockmarked, or otherwise too rough for the dampproofing to properly cover.

Hot-applied bituminous dampproofing is only occasionally used in interior locations because of problems involved with transporting hot materials and the danger of operating heating equipment at interior application sites.

Cold-Applied Bituminous Dampproofing

Cold-applied bituminous dampproofing may be either coal-tar or asphalt based. Cold-applied coal-tar dampproofing is not usually used in interior locations because of its distinctive odor, but it is preferred to asphalt materials for exterior use because it has a higher specific gravity. As is true with hot-applied products, cold-applied coal-tar is not compatible with asphalt.

Cold-applied asphalt dampproofing may be either cut-back or emulsion type. Both types are available either in liquid, semi-fibrated, or heavy fibrated forms. Today, the fibrated types are reinforced with nonasbestos fibers. Older installations, however, may contain asbestos. Most cold-applied dampproofing materials contain inorganic fillers. The composition of individual products varies. Where the dampproofing must pass water vapor (breath) emulsion-type bitumen should be used.

Table 3-3 shows the locations and conditions under which each type of cold-applied dampproofing is usually recommended.

As Table 3-3 shows, cold-applied bituminous dampproofing is used in the same exterior below-grade locations as hot-applied dampproofing but is particularly appropriate for small installation where bringing heating equipment to the site is inappropriate or too costly. Using cold-applied dampproofing makes sense even for large installations where field conditions make transporting hot materials from remote preparation sites difficult, and placing heating equipment at the dampproofing location would be dangerous, difficult, or excessively expensive. As is true for hot-applied dampproofing, the Brick Institute of America's (BIA) *Technical Notes* 7 and the National Concrete Masonry Association's "NCMA—TEK 55" agree that cold-applied below-grade dampproofing on masonry should be applied over parging. Liquid cold-applied dampproofing can be applied directly to dense smooth concrete without parging. Rough concrete to receive cold-applied liquid bituminous dampproofing, however, should be parged. Using fibrated cold-applied dampproofing on concrete that is too rough for hot-applied or liquid cold-applied dampproofing may preclude the need for parging. Follow the dampproofing manufacturer's recommendations about parging concrete.

Cold-applied bituminous dampproofing is also used on the interior faces of exterior walls and on the inner face of exterior wythes of cavity walls. BIA's *Technical Notes* 7 says that mortar coatings are more effective for those uses and cautions against using bituminous materials in such locations,

Table 3-3 Cold-Applied Bituminous Dampproofing Uses

Criteria	Asphalt Emulsion			Asphalt Cut-back			Coal-Tar	DP not Recommended
	L	S–F	F	L	S–F	F		
Exterior								
Smooth concrete	X			X			X	
Rough concrete		X	X		X	X		
Very rough concrete			X			X		
Unparged masonry								X
Parged masonry	X	X		X	X		X	
Parged concrete	X	X		X	X		X	
Soil Conditions								
Wet soil								X
Damp soil				X	X	X	X	
Dry soil	X	X	X	X	X	X	X	
Rough backfill								X
Protection board needed								X
Interior								
Smooth masonry	X	X		X	X			
Smooth concrete	X	X		X	X			
Rough concrete			X			X		
Rough masonry			X			X		
Outside of insulation	X	X	X					
Other								
Breathing required	X	X	X					
Breathing not required				X	X	X	X	
Solvents prohibited	X	X	X					
Damp substrates	X	X	X					

DP = Dampproofing; L = Liquid, S–F = Semi-Fibrated; F = Fibrated

saying that they are not effective, but does not flatly advise against such use. BIA's reluctance to endorse using bituminous materials against interior faces of exterior walls stems partly from the fact that such use permits water to enter the wall but not pass completely through, which leaves the water within the masonry where it can contribute to efflorescence or cause damage when it freezes.

"NCMA—TEK 55" recommends using an impermeable barrier on the interior face of exterior concrete unit masonry walls. A nonbreathing cut-back asphalt dampproofing product might do that job but would have to be concealed and protected by a finish. When an impermeable barrier is used on the interior, the exterior must not be coated with a nonbreathing material, not even oil based or alkyd paint.

Cold-applied bituminous dampproofing is also used for coating structural steel subject to high moisture conditions.

Liquid cold-applied bituminous dampproofing materials are suitable for brush or spray application on hard, smooth, dense surfaces not in contact with coarse backfill materials.

Semi-fibrated cold-applied dampproofing materials are suitable for brush or spray application over rough surfaces.

Heavy fibrated cold-applied dampproofing is suitable for trowel application on coarse substrates such as honeycombed concrete. Any surface too rough for heavy fibrated cold-applied dampproofing should be repaired or parged with a cementitious coating. Parged surfaces will probably be suitable for semi-fibrated bituminous dampproofing and may be suitable for liquid dampproofing.

Bituminous Dampproofing Problems and How to Solve Them

Bituminous dampproofing failures occur for several reasons. Chief among them are the substrate cracks, dislodges, or otherwise fails; waterproofing should have been used in the first place; the wrong dampproofing material was used; the dampproofing material was incorrectly installed; the wall was coated on both sides with impermeable materials, preventing release of built-up moisture.

Bituminous dampproofing will not bridge even hairline cracks. So, when the substrate cracks, the dampproofing will not remain effective. It is necessary to completely and properly repair failed substrates before attempting to repair the dampproofing.

Most bituminous dampproofing failures are due to improper design or installation. One type of improper design results from erroneous assumptions about groundwater. Bituminous dampproofing should never be relied on to exclude even a slight amount of water under hydrostatic pressure. Even materials that might withstand hydrostatic head under some conditions are generally applied too thinly when used as dampproofing to withstand any appreciable pressure. Some dampproofing products will not retain free water under even the slightest pressure.

Water penetrating a wall is a strong indication that hydrostatic head is present, in which case the wall should have been waterproofed instead of dampproofed. When hydrostatic head is present, repairing the dampproofing, adding cants, reinforcing substrate joints, and even applying additional dampproofing coats will not stop water intrusion until the water under pressure is removed. It might be possible to apply a drainage medium to remove the hydrostatic pressure, but doing so would require uncovering the affected walls. It makes no sense to spend the money to excavate the soil covering a leaking wall and then backfill without waterproofing the wall.

Wet or wet-appearing walls, when there is no hydrostatic pressure against the wall, may indicate that the bituminous dampproofing is not functioning properly. Among the possible causes are insufficient thickness of dampproofing, which usually results from applying only a single coat where two or more coats should have been used; failure to install cants at the bottom of walls and turn the dampproofing out onto the footings; failure to reinforce corners and joints in the substrates; improper installation of the dampproofing material; or installation of dampproofing over an incompatible material.

Failures due to coating both sides of a wall with impermeable materials include crazing or spalling masonry faces and blistering of the finish.

Extending and Repairing

Before deciding to repair bituminous dampproofing, get an expert's opinion and recommendations. Repairs below grade can be very expensive, and, if improperly done, or not appropriate in the first place, can be ineffective. In fact, except when the substrate fails, it seldom makes sense to try to repair bituminous dampproofing under any circumstances. Most of the time, the failure means that more protection was needed in the first place and ought to be provided.

When dampproofed walls are extended, extension of the dampproofing will almost certainly be required. When an expansion joint is installed between the original wall and the extension, treatment of the joint must be compatible with both existing and new dampproofing, but the two dampproofing materials do not necessarily have to be compatible. Where the two materials come into contact, compatibility is essential. The best way to ensure compatibility is to have the manufacturer of the new material make compatibility tests at the building site. Specifications should, of course, require that new materials be compatible with substrates and existing materials.

When dampproofing repair is necessary, first ascertain the dampproofing materials originally used and use the same materials when making repairs. Again, have the dampproofing material manufacturer make field compatibility tests, even when the same material is being used. Formulations may have changed. The material in place may have undergone unanticipated chemical changes.

After ensuring compatibility, install the new dampproofing as though the existing material did not exist, with all corner and intersection treatment, reinforcement, and coats recommended by the bituminous dampproofing material manufacturer for new installations. Where substrate repairs were made, make sure that substrate surfaces are consolidated, properly cured (if the dampproofing material manufacturer recommends it), and smooth before making dampproofing repairs. Place the type of reinforcement recommended by the dampproofing manufacturer over small repairs in substrates and at junctures between large repairs and existing undisturbed substrates. Lap dampproofing repairs and patches substantially over existing dampproofing.

When damage results from coating both sides of an above-grade wall with impermeable coatings, the only solution is to remove one or both of the coatings and provide an interior coating that is impermeable and an exterior coating that is not. In some southern climates in air-conditioned buildings, it may be necessary to make the exterior coating the impermeable one.

Clear Water Repellent Coating Types and Their Uses

At first glance, sorting through the many different clear water repellent coating products on the market seems a difficult, if not impossible, task. Some manufacturers state clearly the contents of their product. Others consider themselves proprietary and will not disclose the nature of the materials they use. Even those that list the basis for their product will disclose it to be one of a bewildering array of possibilities, including silicon, methacrylate, metal soap (stearate), silicate ester, titanate ester, gum wax, resin, rubber, butyl, oil, and several other polymers.

Some clear water repellent coating products use a single component as their base. Others are blends. One product might be a methyl methacrylate polymer, for example, while another is a methyl methacrylate and ethyl acrylate copolymer.

Sometimes it is difficult for a nonchemist to determine from their names which category some products fit into. Silicons, for example, include silicone resin (alkylpolysiloxanes, polysiloxanes), silane (alkylakoxysilanes), siloxanes (oligomerous alkylalkoxysiloxanes), and polymeric alkylalkoxysiloxanes. Silicone products may be a single silicon, blends of various silicones, or a silicone combined with other materials such as stearates, silicate esters, or methacrylates. The various types of silicon products are all made from the same raw material—C1-silane—and all result in essentially the same coating material—a polysiloxane or silicone resin. The molecule size may differ, however, depending on which silicone was used; and performance may vary. Some silicons produce heavier coatings than others. Some are easier to apply. Some have solvents that evaporate quickly, while the solvents in others evaporate more slowly. Each producer claims a superior product, of course.

It would be nearly impossible for most of us to find our way around in this chemical morass, except for the fact that most clear water repellent coatings in general use today as wall coatings, including those whose manufacturers will not disclose their contents, are either silicones or acrylics. Even so, selection is not easy. The actual chemical formulation of a product can affect its performance. One methyl methacrylate and ethyl acrylate copolymer may work very well and another product with the identical chemical name may be a poor performer. A product that performs excellently this year may perform poorly next year because of a slight change in its chemistry. Fortunately, there are some industry performance standards to help people who are not chemists select clear water repellent coating products intelligently.

As if selecting clear water repellent coatings for walls was not complex enough, clear water repellent coatings for use on traffic-bearing concrete surfaces offer even more possible formulations. They include boiled linseed oil, epoxies, epoxies containing polysulfide, methacrylate, silane, urethane, butadiene chlorinated rubber, silicate, siliconate, siloxane, isobutylene, and aluminum stearate. Some of them have the same names as the wall materials but are actually different formulations. Fortunately, there is an industry standard to help in selecting them.

One common trait that all clear water repellent coatings should have is to permit virtually unimpeded water vapor transmission. Any product that does not breathe freely should be rejected regardless of other considerations.

Applicable Standards

Selection of a water repellent coating material should be predicated on the product's performance characteristics.

ASTM Standard E 514 presents an industry-recognized test method for determining water repellence of clear water repellent coatings used on brick and block walls. The ASTM Standard E 514 procedure calls for comparing an uncoated wall with the same coated wall and reporting the result of each test (before repellent use and after) in four categories: (1) time to visible moisture; (2) time to visible water; (3) 24-hour rate of leakage; and (4) maximum rate of leakage.

Clear water repellent coating manufacturers should be able to provide results of recent tests based on ASTM Standard E 154 or an equivalent standard. When reviewing product literature, bear in mind that any test results are invalid if the product's formulation has been changed since the test was made.

The following conditions can affect test results:

- The kinds of walls tested. Whether the tested walls were single- or multiple-wythe is significant because ASTM Standard E 514 requires single-wythe walls. Tests run on multiple-wythe walls do not comply with ASTM Standard E 514 criteria and cannot be evaluated according to that standard. Masonry walls with more than one wythe will exhibit more water penetration resistance than will single-wythe walls both before and after coating. The composition of the collar joint in multiple-wythe walls can drastically affect penetration results.

- The type of mortar used. Using masonry cement mortar in the tested assembly will skewer the test because masonry cement mortar is more porous than is Portland cement mortar. If the original mortar is more porous, the difference may appear greater.

- The test methods followed. The product test must follow the exact temperature and humidity conditions required by ASTM Standard E 514 or the equivalent test procedure followed. Minor differences may be exaggerated in the results.

The moisture vapor transmission rate (MVTR) of an installed water repellent coating is just as important as its water repellence characteristics. The major objection of the Brick Institute of America, the International Masonry Institute, the National Building Granite Quarries Association, and the Marble Institute of America to using clear water repellent coatings on clay masonry, granite, and marble is that many water repellent coatings do not breathe. Several test methods are used in the industry to measure MVTR. Those using ASTM Standard D 1653 report results in grams per square foot over 24 hours or a percentage compared with an uncoated sample. National Bureau of Standards Technical Note 883 (see Clark in the Bibliography) recommends a minumum of 20 percent MVTR. Manufacturers using ASTM Standard E 96 report results in percent of weight of water gain.

Portland Cement Association (PCA) Development Department Bulletin D137 (see Litvin in the Bibliography) is a report of PCA Research and Development Laboratories test of sixty clear water repellent coatings products on architectural concrete panels, including exposed aggregate panels and smooth panels. It rates the tested coatings against each other for their ability to protect the concrete against discoloration due to atmospheric contaminates. Trade names are not used, but composition (methyl and butyl methacrylate copolymer, for example) of the tested coatings is. Lack of trade names makes evaluation of products using D137 more difficult, but trade names would be almost useless anyway, because the tests were conducted in 1968. A significant disclosure of the reported tests is the fact that different products containing the same chemicals performed quite differently. One polyurethane resin, for example, was rated ''fair'' while another was "very poor." D137 is still a helpful guide, however, because of the remarks column in the ratings schedules. The "fair" polyurethane resin, for example, is noted as giving a glossy finish, as is the "very poor" material. That is helpful information to have, and suggests a point of query to a manufacturer offering a clear water repellent containing polyurethane resin. D137 also offers a few general conclusions based on the tests. One conclusion, for example, is that many products containing the methyl methacrylate form of acrylic resin performed well. Remember, though, that the tests reported in D137 represent only one criterion for selecting a clear water repellent.

The industry standard for clear water repellents used on traffic-bearing concrete is the National Cooperative Highway Research Program (NCHRP) Report 244 (see Wiss, Janney, Elstner and Associates in the Bibliography). The report includes the results of tests on twenty-one products identified by their generic characteristics. It does not list product names. Any product manufacturer, however, is free to have its product tested according to the methods used in the NCHRP Report 244, and some have.

One word of caution, when comparing a product with the results in NCHRP Report 244, product chemistry can affect performance. A product can fit the generic descriptions in Report 244 and not perform as Report 244 indicates. The results might be similar, but the only way to know is to conduct a test like those reported in Report 244 on the specific product in question. In addition, the tests reported in Report 244 were conducted in 1981. There may be whole classes of products available today that did not exist then.

The same is true when comparing a product with the results of the test reported in Portland Cement Association Development Department Bulletin D137, mentioned earlier. Even though the tests reported in D137 were conducted in 1968, there is no reason to believe that modern materials with the same basic ingredients would not perform similarly; but the only way to be sure is to conduct the same kind of test using the product in question.

NCHRP Report 244 also offers conclusions based on the tests performed. One conclusion is that linseed oil, a widely used product for preventing deicing chemical intrusion into concrete, is not very effective, reducing the chloride content of treated concrete to only 11 percent less than that of uncoated concrete. Some coatings are judged as being slippery and dangerous when used on driving or walking surfaces. Only a few coatings are actually penetrants. A complete reading of Report 244 is helpful when evaluating products for use on traffic bearing concrete.

Field testing of a water repellent coating's coverage rate is essential before entire surfaces are coated. Low coverage rates compared with the material's listed coverage rates suggest that the coating may not be effective as a water repellent coating in that instance. Sometimes, low coverage rates can be overcome by applying additional coats. Too much coating material, however, might form a film which may excessively darken the surface or develop a noticeable sheen, and may not breathe.

Water repellent coatings should be warranted against water penetration for at least 10 years. Product manufacturers who will not warrant their products may be producing a product that will erode quickly in the weather and need recoating often. Such products may not work well even when initially installed.

Some Producers' Associations Recommendations

Advice is available from the producers of the materials that will receive the water repellent coatings, and from the associations that represent them.

Concrete. The Portland Cement Association (PCA) favors using clear water repellent coatings on concrete walls to prevent discoloration due to water intrusion or dirt, to make the concrete easier to clean, and to brighten the color of architectural concrete.

PCA also recommends using clear water repellent coatings on horizontal concrete exposed to water and ice-melting chemicals to protect the concrete against spalling, exfoliation, and other damage. Figure 3-1 shows such an application.

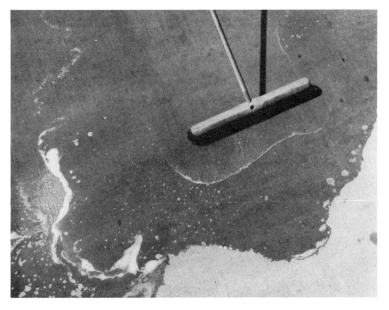

Figure 3-1 Hydrozo Clear 30M application.

PCA discourages using coatings that form a film, or do not breathe, since they trap moisture beneath them which can freeze and damage the concrete.

Silicone and some other products may cause staining in exposed aggregate panels. Review PCA Development Department Bulletin D137 (under Litvin in the Bibliography), and check with the individual panel producer for their recommendations about water repellent coatings.

Brick. The Brick Institute of America's (BIA) *Technical Notes on Brick Construction,* "7E—Colorless Coatings for Brick Masonry," indicates the possible harm from improper use of clear water repellent coatings on brick surfaces. It boils down to some things known in the industry for years. Nonbreathing coatings on masonry can trap moisture in the wall, which causes efflorescence and, in freezing weather, may cause the masonry to spall. Even breathing coatings will not stop water penetration through bad joints or cracks in the masonry. Some coatings, such as silicone, can prevent other coatings from bonding and make treated masonry joints impossible to repoint. Coatings do not prevent efflorescence or staining from occurring and may make them impossible to remove. Coatings applied heavily enough to prohibit graffiti from adhering will probably form a nonbreathing film.

A disadvantage, not mentioned in BIA *Technical Notes* "7E," is that some water repellent coatings, including but not necessarily limited to silicones, tend to wear unevenly, permitting parts of the coated wall to absorb moisture while other portions remain dry. This results in a splotchy appearance that is difficult or impossible to eliminate.

The International Masonry Institute (IMI) has no official opinion about using clear water repellent coatings on clay masonry. The author, however, spoke with an IMI representative who recommended not using them, for several reasons. Manufacturers vary their formulations from time to time to the extent that a product that tests well one year does poorly the following year. The fact that considerable expertise is required to properly apply clear water repellent coatings is not always appreciated. Often, sufficiently experienced applicators are not even available. So, improper application is frequent. Water repellent coatings are also often misused. It is not unusual to find them applied in a vain attempt to stop water from leaking through cracked mortar or broken flashing. The worst example of misuse the author knows about did not come from the IMI representative, however. It is a case in which a misinformed homeowner applied gallons of a nationally known clear water repellent coating to a wall in a location subject to frequent flooding to prevent the wall from leaking. Not only did the wall continue to leak but the solvents in the coating so permeated the house that it had to be abandoned.

IMI might comment on specific products when their opinion is requested but will probably suggest that the caller examine the coating manufacturer's test data and warranties as a basis for selection.

Concrete Unit Masonry.   Both the International Masonry Institute (IMI) and the National Concrete Masonry Association (NCMA) approve of using clear water repellent coatings on concrete unit masonry. "NCMA-TEK 10A" and "TEK 55" recommend where clear coatings should be used and surface preparation methods. "TEK 10A" contains an outdated table comparing types of coatings, most of which are decorative types beyond the scope of this book. Though "TEK 10A" and "TEK 55" do not include all available modern materials, much of their information is worth review by anyone contemplating using a coating on concrete unit masonry. Figure 3-2 shows a clear water repellent coating being applied on a concrete unit masonry wall.

Stone.   The Indiana Limestone Institute (ILI) recommends using clear water repellent coatings on limestone to preserve the stone's clean look and to keep it looking dry in wet weather. ILI does not state which products, or even which classes of products, to use but does offer to comment when queried. Silicons, which can combine with iron and some other materials in limestone and make unsightly stains, probably should not be used without

Figure 3-2 Hydrozo Clear Double 7 application.

ILI's concurrence. Consult with the ILI before using any particular product or water repellent class.

The Marble Institute of America (MIA) has no published recommendations about putting clear water repellent coatings on marble but generally disagrees with the practice. A representative says, however, that while some MIA members think that using clear water repellents on marble is a bad idea, not all members agree with that conclusion.

The Building Stone Institute (BSI) recommends that slate never be coated with clear water repellents regardless of the location of the slate.

The National Building Granite Quarries Association (NBGQA) recommends that clear water repellent coatings never be used on granite under any circumstances.

Contact the individual producers of other types of stone to determine their recommendations about using water repellent coatings. Usually, hard, dense stone should probably not receive water repellent coatings. Some water repellent coatings may help some porous stone types maintain their appearance. Others may cause unsightly stains. Silicones, for example, react with iron and may damage iron-bearing stone.

Clear Water Repellent Coating Problems and How to Solve Them

Virtually every association representing producers of products to be coated with water repellent coatings recommends that the cause of problems be fully determined before a decision to use water repellent coatings is made. Indications that might lead one to assume that a water repellent coating was appropriate are wet walls, leaks, condensation, efflorescence, staining, and spalling of concrete. In every case, the first order of procedure is to determine the cause of the damage.

Every clear water repellent coating used should breathe. Some coatings will preclude later coating with another coating material or paint. If such future treatment is a possibility, use a clear water repellent coating that will not block future action.

Always follow the coating manufacturer's instructions completely when installing water repellent coatings, including its recommendations for preparing the surface, cleaning the surfaces to be coated, and mixing and installing the coating. Ensure that the installer is experienced. Many problems with water repellent coatings result from improper installation.

Before coating installation, have the new coating manufacturer conduct field tests to verify compatibility of the new product with the existing one and with all other materials which will come into contact with the new water repellent coating.

Traffic-Bearing Concrete

For horizontal concrete showing damage from chloride ion intrusion, the first step is to determine whether the damage has already grown severe enough to require major repair or replacement of the concrete. The decision should be made only with the recommendations of an expert in the field, such as a structural engineer.

Where the damage is not severe enough to require major repairs, repair the affected concrete surfaces. If the concrete has been previously treated with a water repellent coating, ascertain the name of the product used and select a product that will be compatible. It may be necessary to use the same product to achieve compatibility, which there may be a reluctance to do, especially if the original product did not perform as expected.

Walls

Visible water-related damage or unsightly conditions on wall surfaces are not usually caused by water passing through the wall material itself. Probable causes vary with the wall material and construction.

Concrete.   Damage or unsightly conditions may result from condensation or leaks. Exposed concrete may be soiled or stained or may darken when wet. The Portland Cement Association (PCA) recommends ascertaining and correcting the cause of the damage or unsightly condition. The cause of condensation (see Chapter 2) should be removed, the leaks repaired, and the concrete thoroughly cleaned before a decision is made to use a water repellent coating. If a coating is necessary, select it based on PCA recommendations, referenced standards, such as Albert Litvin's PCA Development Department Bulletin D137, "Coatings for Exposed Architectural Concrete," and the recommendations of the coating manufacturer.

Brick.   Wet-looking brick masonry walls, efflorescence, and spalling bricks are usually caused by leaks through joints or flashing. Clay brick is a relatively impervious material. Glazed brick is as impervious as ceramic tile.

The Brick Institute of America's *Technical Notes* 7E includes a ten-point detailed checklist for investigating the cause of visible damage and the installation of clear water repellent coatings. Following that checklist will not ensure that a water repellent coating on brick masonry will not cause problems later, but failure to follow it will almost certainly guarantee failure. The checklist is not repeated here because it would be superfluous. Anyone with the kinds of problems that would lead them to try a clear water repellent coating on a brick wall should read every word the Brick

Institute of America has to say on that and related subjects before taking any action anyway.

The International Masonry Institute (IMI) recommends looking for other causes before considering the use of water repellent coatings on clay masonry and then not using the water repellent coating. But if someone really insists on using a water repellent coating on a brick wall, IMI will comment, when queried, on their past experiences with various specific coating materials.

Concrete Unit Masonry.   Before considering clear water repellent coatings on concrete unit masonry walls that display efflorescence, leaks, spalling, or other obvious water damage, or fungal growth suggesting wetness, check to see if there is a leak. Leaks can result from failed joints or flashings or cracks in the unit masonry itself. Also check to see if the wetness or water damage is caused by condensation. Condensation is frequent in single-thickness walls and uninsulated walls.

If after eliminating all possible causes of leaks and condensation a wet condition persists, consider using a water repellent coating.

Unprotected concrete masonry units are highly porous and will tend to absorb water directly through their faces. Exterior concrete masonry unit walls that are not coated with decorative materials or paint are sometimes given a clear water repellent coating when they are new.

Where leaks or condensation are not present, concrete unit masonry walls that have not been previously coated or painted and which appear wet after a rain are candidates for a water repellent coating. If the original masonry's natural appearance is desirable, a clear coating is appropriate.

Concrete masonry units that have been previously given a clear water repellent coating may appear splotchy or wet in some places and dry in others. That is probably a result of the original coating wearing off unevenly. Recoating is appropriate. Use the same material that was originally used, if it did a satisfactory job for the expected life of the product. Otherwise, use a different product that is compatible with the existing material.

Stone.   Efflorescence and visible water that penetrates through stone walls are usually the result of damaged stone, failed joints, or improperly installed or broken flashings. That is true even when the stone is limestone or another highly porous stone. Always verify that the condition is not the result of a leak before considering using a water repellent coating.

A wet appearance which continues after all leaks have been repaired may indicate that water is intruding through the face of the stone. When the stone is granite, slate, or another relatively impervious hard stone, the condition should probably not be treated. Consultation with the installer's or producer's representative may result in a different conclusion, but the condition is one that will probably not be correctable. Treatment with a water repellent coating, in the face of the producer's recommendations to the contrary, will probably not alleviate the problem and may result in damage to the stone through freeze-thaw action on moisture that will be then trapped in the wall.

Fortunately, very hard stones, such as granite and slate, have low absorption rates, which prohibit excessive water penetration.

It might be permissible to treat wet-appearing marble with a water repellent coating after all possible leak sources have been repaired. The Marble Institute of America recommends against using water repellent coatings on marble but concedes that its opinion is not universally shared in the industry, even by its own members. If wet-appearing marble is a major problem, contact the stone producer and water repellent coating product producers for their advice. It is best to follow consensus advice. A wet-appearing wall after a rainstorm is preferable to spalling stone.

Limestone and other soft stones will often have received a water repellent coating when new. Coated walls might begin to show a splotchy appearance with some parts appearing wet and others dry. This condition often results from the original coating wearing unevenly. After eliminating all possible leaks, let the wall completely dry out and renew the original coating.

Extending and Repairing

Repairing existing water repellent coatings consists of recoating the surfaces.

Extending existing water repellent coatings is sometimes necessary when originally coated walls are extended. When coatings are extended, it is necessary to ensure that the new coating is the same material as, or at least compatible with, the original. In addition to maintaining chemical compatibility, it may be desirable to use a new coating that produces a finish matching the original. Some coatings, even though advertised as clear, produce a distinctive sheen on some materials and darken some surfaces. Coatings with different chemical bases can produce quite different appearances.

Where to Get More Information

Request specific data, including test reports where appropriate, from bituminous dampproofing and clear water repellent coating materials manufacturers. Manufacturers' names, addresses, telephone numbers, and lists of their products are available in CSI's *Spec Data, Sweet's Catalog File, Masterspec,* and the trade associations representing the manufacturers.

AIA Service Corporation's *Masterspec,* "Basic: Section 07160" and

"Section 07175," address the issues outlined in this chapter related to product selection and use.

Masterspec recommends that editors of "Section 07175, Water Repellents," have Portland Cement Association (PCA) Bulletin D137 (see Litvin in the Bibliography) and National Bureau of Standards (NBS) Technical Note 883 (see Clark in the Bibliography) available for reference. It is a good idea, even for those not using *Masterspec,* to have them available. Contact PCA and NBS for copies.

PCA *Design and Control of Concrete Mixtures, Thirteenth Edition* is a comprehensive reference on the subject of its title but says only a little about coatings, other than that they are often used and that those consisting of acrylic resins in methyl methacrylate form are among the best for use on concrete.

PCA publications "Permeability Tests of Masonry Walls" and "Repairing Damp or Leaky Basements in Homes" are good reference materials for anyone considering clear water repellent coatings for use on concrete or concrete masonry.

PCA publication "Effects of Substances on Concrete and Guide to Protective Treatments" describes twenty-two types of substances that might be used to protect concrete and discusses other coatings and protective treatments such as waterproofing and dampproofing. The publication also includes a table of materials that damage concrete and suggests proper protective materials for each type of damaging material. The last portion of the publication lists manufacturers of products in each of the described twenty-two protective substances types.

National Cooperative Highway Research Program "Report 244" (under Wiss, Janney, Elstner and Associates in the Bibliography) is a report of a research study which evaluated twenty-one products for use on concrete surfaces to prevent intrusion of damaging chloride-laden water. It is an excellent report that should be studied by every professional responsible for selecting or specifying such products. The report is available from the Transportation Research Board of the National Academy of Sciences.

PCA "Removing Stains and Cleaning Concrete Surfaces" offers guidance for cleaning concrete before using clear water repellent coatings.

The National Concrete Masonry Association's "NCMA—TEK 10A" and "NCMA—TEK 55," although somewhat dated, contain NCMA's official position on clear water repellents. "TEK 55" also addresses interior and exterior below-grade dampproofing of concrete unit masonry walls.

The Indiana Limestone Institute's *ILI Handbook* states ILI's position on using clear water repellent coatings on limestone. ILI will comment, when asked, on the qualities of acceptable water repellents.

CHAPTER

4

Waterproofing

The term "waterproofing" in this chapter refers to materials used to waterproof below-grade walls, tunnels, and pits, and horizontal decks below paving or earth. Except for cementitious waterproofing, the waterproofing discussed here is concealed in the completed building. Associated flashings and accessories may, however, be exposed to view. The types of cementitious waterproofing discussed here are intended for use in such utilitarian locations as elevator pits, tunnels, and garages.

The following are not discussed in this chapter: decorative membranes in the category known as "traffic topping," which are intended to be exposed on decks, promenades, and sports courts; waterproofing used between interior spaces, such as beneath supported kitchen floors or shower stalls; flooring products, even though they may act as waterproofing; cementitious products applied to walls above grade, which may block water but are not waterproofing as defined here; chemical waterproofing mixed into poured concrete.

Most applications of waterproofing are on new materials to prevent water under hydrostatic head from entering those materials. New waterproofing is also sometimes applied to existing surfaces. When water penetrates below-grade walls, waterproofing is usually the remedial material choice. Properly selected and installed waterproofing should last for the life of the building, unless it is damaged by some outside force. Improperly selected or installed waterproofing does fail occasionally, forcing repair or replacement. Building additions often require extension of existing waterproofing.

This chapter suggests some guidelines for waterproofing use and offers recommendations for extending existing waterproofing, repairing existing waterproofing, and applying new waterproofing over existing surfaces, including existing waterproofing.

Before discussing waterproofing in detail, it is necessary to show how waterproofing differs from bituminous dampproofing and water repellent coatings, which are discussed in Chapter 3. Table 3-1 shows where dampproofing, water repellent coatings, and waterproofing are likely to be used.

Bituminous dampproofing is a bituminous coating used to prevent building materials from absorbing adjacent moisture. Like waterproofing, bituminous dampproofing is often used on below-grade surfaces. Unlike waterproofing, dampproofing is not intended to keep out water under pressure and is not used on horizontal surfaces. Bituminous dampproofing is also used on the interior surfaces of exterior above-grade walls, on interior surfaces of the exterior wythe of exterior cavity walls, and for coating structural steel subject to high moisture conditions. Waterproofing is not used in any of those locations.

Clear water repellent coatings are used to reduce water penetration into and soiling and staining of exterior wall surfaces above grade. Clear water repellent coatings are also used to prevent damage of exposed horizontal concrete by water, sodium chloride, and other ice-melting chemicals.

Horizontal concrete coatings intended primarily to aid curing are beyond the scope of this book.

What Waterproofing Is and What It Does

Waterproofing is an impervious (relatively) membrane, coating, or sealer used to prevent water from entering or passing through either horizontal or vertical building materials. Most waterproofing either covers surfaces or plugs openings. Typical uses for waterproofing materials are as coatings on exterior walls below grade where occupied space occurs on the interior side of the wall; on the interior surface of walls below grade where the exterior surface is inaccessible; on horizontal deck surfaces below paving or earth fill; and on the walls and roofs of tunnels. Waterproofing sealers are used to plug holes through below-grade walls, sometimes in the face of flowing water.

Waterproofing is designed to exclude water even when the water is under hydrostatic head. Most waterproofing membranes also inhibit water vapor transmission, which under some circumstances can be detrimental. Trapped water vapor can blister some membranes or condense and freeze, damaging the membrane or the substrates.

Cementitious waterproofing can be installed on either the interior or exterior side of a concrete or masonry substrate. All other waterproofing types discussed here must be installed on the same side as the water and be supported from the opposite side.

Waterproofing Types and Their Uses

There are three basic types of waterproofing: membrane, clay, and cementitious.

Requirements Common to All Waterproofing Types

Waterproofing Systems

Membranes Systems. On vertical surfaces, a membrane waterproofing system usually consists of a membrane, a protection course, a drainage medium, and a layer of filter fabric (see Fig. 4-1). The parging shown in Figure 4-1 is used when the substrate is unit masonry.

Horizontal membrane waterproofing systems usually consist of a membrane, a protection course, a drainage medium, a layer of filter fabric, and a wearing surface (see Fig. 4-2).

The protection course may be protection board or insulation. The drainage medium may be either a gravel bed, a pervious board, or a three-dimensional sheet. Unless filter fabric is a component of the drainage medium, a separate layer of filter fabric is needed over the drainage medium.

In horizontal applications, the wearing surface may be pavers in a mortar setting bed, pavers in a sand bed, pavers set without a bed, poured concrete, or soil. Some pavers are grooved on the back side for drainage. Others can be mounted on pedestals. When either is used, the drainage medium can be omitted from the systems. Water penetrating the pavers will run to the drains through the perforations or between the pedestals. The filter fabric is usually recommended, however, even when there is no drainage medium, to prevent small particles from reaching the protection board and membrane. The filter fabric type should be that recommended by the waterproofing system component manufacturers.

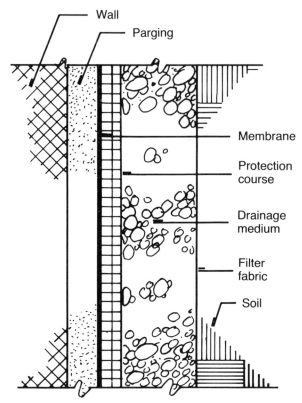

Figure 4-1   Vertical membrane waterproofing system.

Sometimes, the insulation in a waterproofing system is placed beneath the membrane. Then special considerations must be given to prevent condensation from forming in the system, and a separate protection board is necessary.

Bentonite Systems.   A typical bentonite waterproofing system consists of bentonite applied over the substrates. A drainage medium is seldom used. Where it is used, filter fabric is also required. A protection course is used in vertical applications only where the backfill includes sharp rocks. Protection courses are used more frequently when the bentonite is on a horizontal surface, but even there is often omitted.

Cementitious Systems.   Cementitious plaster systems are troweled on in two, three, or four courses. The top course is a hard-finish plaster or a poured-in-place concrete topping to protect the lower courses from damage. Some systems are sprayed on.

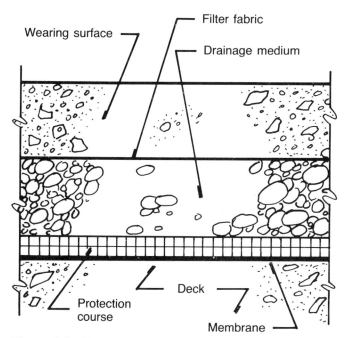

Wearing surface

Filter fabric

Drainage medium

Deck

Protection course

Membrane

Figure 4-2 Horizontal membrane waterproofing system.

Cementitious waterproofing on exterior surfaces is rarely covered with protection board or a drainage medium. When either is used, the principles applicable to using them in membrane systems apply.

Cementitious waterproofing is not usually used in horizontal applications.

Compatibility. Every material used in each waterproofing system must be compatible with all other materials in the system or adjacent to the application.

Substrates. Decks to receive waterproofing should be stable, strong enough to support the loads to be encountered without undue deflection, sloped to drains, free from fins and offsets detrimental to the waterproofing membrane, and smooth. Irregularities exceeding 1/4 inch per foot should be removed.

Deck slope should be at least 1/8 inch and preferably 1/4 inch per foot. The slope should be positive and present even after the structure has deflected.

The best deck for waterproofing is an integrally sloped poured-in-place concrete slab. Toppings are more likely to crack and move differentially, which can damage membranes, and leaks are harder to find when toppings are used, even when the waterproofing is fully adhered, because water can migrate beneath the topping.

Where precast concrete decks must be used, they should receive a continous reinforced concrete topping at least 1-1/2 inches thick.

Some waterproofing membrane manufacturers require that their products not be installed over any material (concrete, tile, pavers, etc.) that was installed using a PVC or latex additive or coating. Others do not have such a restriction. If such a condition exists, it should be pointed out to the new waterproofing manufacturer and advice about it should be followed.

Expansion joints should be installed in the substrates to account for thermal, seismic, and settlement movements. Where joints occur, they should be accommodated in the waterproofing system.

Concrete to receive waterproofing should be cured the number of days required by the waterproofing manufacturer. Regardless of the waterproofing type, concrete substrates should not be cured using chemicals. Required curing time varies depending on the product to be used, but it is often 7 or 14 days.

Surfaces to be waterproofed should be smooth, in plane, clean, dry, and free from dirt, dust, and other foreign substances. Concrete should have a wood float or light troweled finish. Patches necessary to eliminate depressions, honeycomb, pockmarks, tie holes, and other irregularities should be made using epoxy or latex-modified mortars depending on the advice of the waterproofing manufacturer. Some manufacturers recommend using their proprietary patching materials for making repairs (see Fig. 4-3).

Surfaces, such as concrete, masonry, and metal, to receive bituminous or other liquid-applied materials should be primed, unless not recommended by the waterproofing materials manufacturer. Primers should be types and materials recommended by the waterproofing manufacturer. Priming may not be necessary for some waterproofing systems. Substrates for a bituminous membrane system may not need priming, for example, when the bitumen is coal-tar-based.

Wood and other nailable substrates to receive waterproofing should be covered with an asphalt-saturated organic fiber felt base sheet or a slip sheet of rosin-sized building paper. Both should be nailed in place. Asphalt-saturated base sheet should not be used where incompatible with waterproofing membrane materials. The need for separation from the substrates, of course, limits the number of systems that can be used over nailable substrates, especially wood.

Drainage.   Waterproofing systems work better and last longer when adequate drainage is provided to lower hydrostatic head. Most waterproofing materials eventually disintegrate when water is constantly present. Emulsified (water-based) asphalt compounds are particularly susceptible.

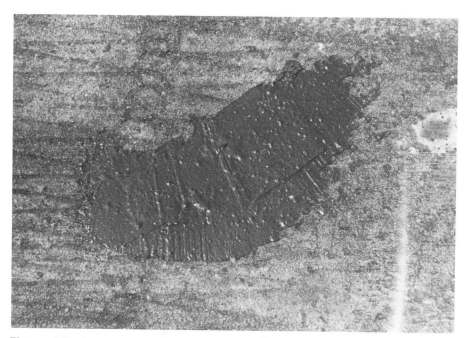

Figure 4-3 Concrete deck damage repaired using W. R. Grace & Co. *LM 3000.* Deck will receive Bituthene sheet waterproofing. (*Photo courtesy of Rick Scruggs, W. R. Grace & Co.*)

Drainage is usually accomplished by placing a layer of pervious material between the waterproofing and the water. Often, the previous layer is continuous over the entire surface of the waterproofing. In vertical applications, pervious chimneys are sometimes used. Whether the drainage medium is continuous or in chimneys, the water drained through it must be carried away from the building. In vertical applications, this is usually accomplished by means of a system of open-joint or perforated pipe drains at the base of the drainage medium. Horizontal surfaces should be sloped to drains. Regardless of the wearing surface material, drains in horizontal surfaces under most circumstances should be of a type designed to receive water at both the traffic surface and membrane levels. Only when the wearing surface is an open-joint type that permits water to flow freely through it should drains be open just at the membrane level.

The traditional drainage medium for both vertical and horizontal applications is gravel. Gravel is still used extensively today, but there are also several types of prefabricated boards and pervious sheet materials manufactured to lead water away from waterproofing. Among them are beaded polystyrene boards, three-dimensional plastic sheets, and glass fiber panels.

Drainage medium should be covered with filter fabric to prevent small-particle (dirt) intrusion into the medium. Some prefabricated drainage products have a factory-applied filter fabric cover on one side. Others do not. Unless filter fabric is part of the drainage medium, a separate filter is always required.

Insulation.   In most waterproofing systems where insulation is required, the insulation is placed on the weather side of the waterproofing layer to protect the waterproofing layer and to lessen the chance of condensation forming below the waterproofing. The insulation generally used in horizontal waterproofing systems where a wearing layer is present is extruded polystyrene (see Chapter 5).

Insulation is sometimes placed below the waterproofing layer. When it is, special consideration should be given to preventing condensation. In addition, placing the insulation below the waterproofing limits the usability and negates the advantageous features of some waterproofing types. One major advantage of fully adhered membrane waterproofing systems, for example, is that they make finding leaks easy. Water cannot migrate between the membrane and the deck. Applying those systems to insulation eliminates that advantage.

Wherever the insulation occurs in the system, compatibility is critical. The insulation and waterproofing materials manufacturers should be consulted about which insulation is appropriate in each particular waterproofing system.

Nailers.   Some, but not all, waterproofing systems require nailers. Where nailers are required, they should be pressure-preservative-treated wood members set with their faces flush with the substrate.

Protection Course.   Waterproofing subject to damage during backfill or wearing surface installation or use should be protected by insulation or protection boards, which covers just about every waterproofing installation (see Fig. 4-4).

Protection boards are asphalt-core composition or fiberboard. Thermal insulation and board-type drainage media, where needed, can also serve as protection board.

Composition protection boards are semirigid sheets composed of asphalt-saturated felt layers. Thicknesses vary from 1/8 inch to 1/2 inch. Usually, vertical applications are 1/8-inch board and horizontal applications are 1/4-inch board, and 1/2-inch board is used in special cases, such as below vehicular traffic topping.

Figure 4-4 Light gray material to the left is installed protection board. To the right, protection board is stored upside down ready for installation on the membrane (dark strip in the middle). (*Photo courtesy of Rick Scruggs, W. R. Grace & Co.*)

Fiberboard protection board is treated, asphalt-saturated and coated organic fiberboard. The thickness is usually 1/2 inch. Fiberboard should not be used to protect horizontal membranes.

Protection courses should be positively fastened to the membrane. In hot-applied systems, the protection course should be applied in the top coat before it cures. In cold-applied systems, the protection course should be applied over dry or cured membrane using mastic compatible with the waterproofing and the protection course material.

Waterstops.   Waterproofing is most subject to failure at joints in the structure. Where hydrostatic pressure is present, waterstops in the substrate's joints are sometimes recommended. Belief in the advisability of using waterstops, however, is not universal. Some structural engineers believe that waterstops actually contribute to water penetration because they are so often deformed during concrete placing, which opens a joint through the wall. Nevertheless, some form of expansion control must be imposed at joints and turns in the substrate and at changes in the substrate material if leaks are to be prevented.

Cant Strips.   Cant strips should be installed at changes in direction in bituminous waterproofing systems, in liquid-applied waterproofing, and in other waterproofing systems where recommended by the waterproofing manufacturer.

Flashings.   Membrane waterproofing membranes are sometimes extended at edges, openings, and projections to act as their own flashings. Rubber or plastic flashings are usually used in liquid-applied waterproofing systems and may be used in other membrane systems if specifically approved by the waterproofing membrane manufacturer. Some membrane flashings should be covered with protective coatings to prevent damage from weather or sunlight.

Bentonite is, for the most part, self-flashing. Joints and edges are sealed using bentonite.

Cementitious materials can only be flashed in the same sense that plaster or any other cementitous materials are flashed. Cementitious waterproofing is brought up to within 1/4 inch of abutting surfaces and the joints are filled with sealant. Tops of cementitious waterproofing subject to water drainage should be capped with metal flashing.

Metal cap flashings are discussed in Chapter 8.

Weather Conditions.   Waterproofing should not be installed during inclement weather or when precipitation threatens.

The ambient temperature when waterproofing is installed should be within the limits recommended by the waterproofing membrane materials manufacturer. Usual minimum temperature without implementing special procedures is 40 degrees Fahrenheit.

Expansion and Contraction.   In addition to provisions in the substrates for building movement, expansion and contraction in the membrane must be provided for. Manufacturers' instructions should be followed. Moving joints in the substrate should continue through the waterproofing with appropriate treatment to prevent water from entering the joint.

Membrane Waterproofing Materials

Membrane waterproofing materials include felt and bitumen membranes, plastic and rubber sheet membranes, and liquid-applied membranes.

Bitumens. Both coal-tar and asphalt bitumens are used in membrane waterproofing. Suitable asphalt bitumens are classified in ASTM Standard D 449 as Type A and called "self-healing." Suitable coal-tar bitumen is classified in ASTM Standard D 450 as Type II, and called "waterproofing pitch."

The bitumen temperature requirements listed in Chapter 7 are applicable to bitumens used in waterproofing.

Plastic Cement, Adhesives, Primers, and Coatings. Plastic cement used in waterproofing systems is the same as that used in built-up bituminous roofing applications. Refer to Chapter 7 under "Built-Up Bituminous Membrane Roofing" for a discussion about plastic cement.

Different adhesives and primers are necessary to install and seal the joints in each different membrane waterproofing material. Only adhesives and primers furnished or recommended by the membrane manufacturer should be used.

Asphalt emulsion should comply with ASTM Standard D 1187.

Asphalt mastic should comply with ASTM Standard D 491.

Oil-based paints are not appropriate over waterproofing or associated flashing membranes unless the manufacturer specifically approves.

Bituminous Felts and Building Paper. Bituminous felts and building paper used in waterproofing are the same as those used in built-up bituminous roofing. Refer to Chapter 7 under "Built-Up Bituminous Membrane Roofing" for a discussion of various types of felt.

Waterproofing Sheets. There are many waterproofing membrane sheet types available. Even though an attempt has been made to include the major types in use at the time of this writing, the following list may not be complete when you read it.

The materials listed are either factory-cured (vulcanized) elastomers, uncured (nonvulcanized) elastomers, thermoplastics, or polymer-modified bitumen. Refer to the detailed discussion in Chapter 7 of the meaning and application of these terms.

EPDM, Neoprene, CSPE (Hypalon), CPE (Polyethylene), PVC, and Polymer-Modified Bitumen. Refer to Chapter 7 under "Single-Ply Membrane Roofing" for a detailed description of these materials. Waterproofing membranes made from these materials are generically the same as roofing membranes, although some differ slightly in formulation. Waterproofing membranes of these types tend to be on the thicker end of the thickness available in each case. Modified bitumen sheets used for waterproofing are sometimes called "rubberized asphalt" sheets. Some waterproofing sheets normally containing plasticizers are blended with other polymers to eliminate or reduce the plasticizer content of the product and help eliminate membrane failure caused by plasticizer migration.

Butyl. Nonvulcanized synthetic rubber sheets. Some butyl sheets are blended with EPDM. Butyl waterproofing sheets are usually 1/16 inch thick.

Preformed Waterproofing Membrane. A composite material made by bonding layers of bituminous saturated felts and fabrics together with bitumen, coating each side with rubberized asphalt or coal-tar pitch, and covering one or both sides with a film of paper, polyethylene, PVC, or similar nonstick material. Preformed waterproofing membranes usually weigh about 60 pounds per 100 square feet.

Liquid Membrane Materials.   Many liquid-applied waterproofing membrane materials are available. The most commonly used today are one-part urethane, two-part urethane, and hot-applied rubberized asphalt. Polysulfide systems are not manufactured today as standard products but may be found on existing structures. Cold-applied asphalt liquid waterproofing is not widely used but is available and may also be encountered in an existing building. Neoprene, Hypalon, epoxy and silicone-based systems might also be found in concealed existing waterproofing, but those materials are usually used in exposed applications, which are beyond the scope of this book. Other products which might be encountered are rubber, polyester, and materials made by blending the mentioned materials with bituminous products.

Bituminous Membrane Waterproofing Installation

Requirements Common to All Bituminous Membrane Waterproofing. The Brick Institute of America (BIA) and the National Concrete Masonry Association (NCMA) recommend that bituminous waterproofing, where appropriate on unit masonry walls, be applied over cementitious parging.

The Portland Cement Association (PCA) recommends parging on concrete only when the wall is honeycombed, pockmarked, or otherwise too rough for the waterproofing to properly cover.

Bituminous membrane waterproofing is installed in plies similar to built-up bituminous roofing. The number of plies varies depending on the hydrostatic head and whether the membrane is a hot-applied or cold-applied type.

In horizontal applications, systems with plies of felt or fabric may be installed in shingle fashion with the laps in the direction of water flow. In horizontal applications where both felts and fabric are used, and in vertical applications, plies should be installed using the so-called waterproofing method.

In the shingle method, all plies are rolled out simultaneously, each lapping the previous.

In the waterproofing method each ply is installed over the entire surface. Subsequent plies are then installed over the preceding ply, also covering the entire surface. Succeeding plies should lap over previous plies by about 3 inches. Where alternate plies are fabric and felt, one layer of each may be combined and laid as one ply, or each ply may be laid separately.

Special precautions should be used when applying bituminous membrane waterproofing on vertical surfaces. Unless the manufacturer specifically recommends otherwise, plies should be placed vertically in lengths of from 6 to 8 feet. The top of each course should be nailed in place at about 8 inches on center. End joints should be staggered from those in underlying plies.

Corners and areas of stress should be reinforced with two plies of glass fiber or cotton fabric before the other plies in the regular system are installed. The reinforcing plies should be set in plastic cement or bitumen. At corners and other locations where membrane joins another membrane the two pieces should be lapped at least 10 inches.

Cut-offs should be installed when work stops in incomplete membranes to prevent water from entering the system.

Before membrane waterproofing on horizontal surfaces is covered, the installed membrane should be flood-tested by damming and flooding with water for 24 hours. Observed leaks should be repaired and flood-tested again until no leaks appear. Where flood testing is not possible, spray testing and other testing methods may be used.

Hot-Applied Bituminous Membrane Waterproofing Application. Hot-applied waterproofing is a commonly used bituminous waterproofing. It works quite well on horizontal decks. Although difficult to apply there, and not as successful as one would hope, it is also used vertically below grade on exterior building walls where occupied space occurs on the interior side of the wall.

Hot-applied bituminous waterproofing may be either coal-tar-based or asphalt-based, but coal-tar is preferable. Coal-tar and asphalt are not compatible and should not be used in the same application. Felts may be either glass fiber or organic-saturated felt, usually No. 15. Fabrics are usually bitumen-saturated cotton or woven burlap. Treated glass fiber fabric is sometimes used where extra strength is desired.

Plies of felt and fabric are alternated in most systems. The top ply is usually felt.

The total number of plies (both felt and fabric included) usually used in hot-applied bituminous waterproofing systems is two for a hydrostatic head not exceeding 3 feet, three for a hydrostatic head between 4 and 10 feet, four for 11 to 25 feet, five for 26 to 50 feet, and six for 51 to 100 feet. There should be one more bitumen layer than ply, because the entire surface of the installed membrane should be coated with a layer of bitumen. The top ply should not be exposed.

Cold-Applied Bituminous Membrane Waterproofing Application. Cold-applied bituminous waterproofing membrane is a built-up membrane consisting of glass fiber fabric plies with asphalt mastic or asphalt emulsion coats. Asphalt emulsion must be allowed to cure completely before coming into contact with water to prevent reemulsification.

Some authorities believe that use of cold-applied asphalt membrane waterproofing should be limited to cases where hydrostatic head is 3 feet or less. Installations have been made, however, where hydrostatic heads were higher. Cold-applied asphalt membranes should have two plies for a hydrostatic head not exceeding 3 feet, three plies for a hydrostatic head between 4 and 15 feet, and four plies for a head between 16 and 30 feet. There should be one more asphalt mastic or asphalt emulsion layer than ply, because the entire surface of the installed membrane should be coated with a layer of bitumen. The top ply should not be exposed.

Sheet Membrane Waterproofing Installation

Each sheet membrane type requires special installation procedures specific to that type (see Fig. 4-5). The manufacturer's recommendations should be followed unless they are obviously incorrect. There are, however, some general requirements applicable to all.

Sheet membranes must be continuously supported to prevent punctures.

Very thin materials (4-mil polyethylene, for example) are not waterproofing and should not be used for that purpose. Usual thicknesses for waterproofing membrane sheets are 1/16 inch, 56 mils, and 60 mils.

On horizontal applications, the largest practicable sheets should be used. On vertical surfaces, however, sheet weight must be considered. There, sheets that are too wide may not be able to support themselves against gravity. Joints must be carefully handled in vertical applications, also, because the sheets will stretch under gravity and deform or break the seal in improperly made joints.

Sheets should be fully adhered or partially adhered according to the manufacturer's instructions. Sheets are not usually bonded to moving substrate joints.

Figure 4-5 Waterproofing sheets (Bituthene) laid stairstep fashion over primed concrete deck. (*Photo courtesy of Rick Scruggs, W. R. Grace & Co.*)

Joints in sheets should be lapped (usually 3 inches) and sealed according to the manufacturer's instructions.

Sheet edges should be sealed using the manufacturer's recommended adhesives and sealants.

Penetrations should be the minimum number possible. The more penetrations through a membrane, the higher the risk of leaks.

Exposed sheets should be resistant to sun and weather or should be coated with materials that are resistant.

Before sheet membrane waterproofing on horizontal surfaces is covered, the installed membrane should be flood-tested by damming and flooding with water for 24 hours. Observed leaks should be repaired and flood-tested again until no leaks appear. Where flood testing is not possible, spray testing and other testing methods may be used.

Liquid-Applied Membrane Waterproofing Installation

Each liquid-applied waterproofing membrane product requires special installation procedures specific to that product. The manufacturer's recommendations should be followed unless they are obviously incorrect. There are, however, some generic requirements.

Some systems require glass fiber cloth or mesh reinforcement throughout. Others require reinforcement only at corners, intersections, and visible cracks. In every case, the membrane thickness should be doubled over cracks, and at corners and intersections, even when there is reinforcement.

Waterproofing membranes should be separated from moving substrate joints by a slip sheet. Some manufacturers recommend covering expansion joints with sheet rubber stripped into and covered by the liquid-applied system. Others require the waterproofing system to be interrupted at expansion joints.

Final dry film thickness varies from product to product but is very important for the system to function properly. Proper application is a major problem with fluid-applied systems. Experienced personnel and care during application are a necessity. Many products shrink during curing, which makes necessary an application thicker than the final desired dry film thickness. The National Roofing Contractors Association recommends that vertical liquid-applied waterproofing membranes be not less than 60 mils thick and that horizontal applications be at least 75 mils thick. These recommendations exceed many manufacturers' recommendations.

Before liquid-applied waterproofing on horizontal surfaces is covered, the installed membrane should be flood-tested by damming and flooding with water for 24 hours. Observed leaks should be repaired and flood-tested again until no leaks appear. Where flood testing is not possible, spray testing and other testing methods may be used.

Bentonite Waterproofing

Material.   Sodium bentonite is an expanding clay material. When wetted, it expands to between ten and fifteen times its dry volume, filling gaps and welding itself together into an impervious mat. Bentonite is available in dry sheets between cardboard, adhered to a drainage board, and adhered to a plastic sheet. Bentonite can also be mixed with water and sprayed in place.

Application.   Bentonite waterproofing is used below grade in concealed locations, on the exterior surfaces of walls, over tunnels and other structures beneath earth fill, and beneath slabs-on-grade. It should not be installed on a supported horizontal structure beneath rigid wear surfaces or any material less than 18 inches thick. Sometimes, manufacturers recommend even deeper cover.

Bentonite is often applied to shoring and sheeting or against existing construction before new walls are placed, where access for application of membrane materials is not available (see Fig. 4-6). Bentonite can also be installed after walls are built. Complete cure of concrete is not necessary before bentonite application, but some cure will prevent excess swelling. Masonry to receive bentonite should be covered with at least 1/2 inch of cement plaster or parging. Bentonite can only bridge cracks up to 1/8-inch wide. Wider cracks must be filled.

Bentonite thickness should be doubled over joints in the substrate and at corners and intersections. Sometimes, bentonite strips are embedded in concrete joints as added insurance against water penetration.

Sometimes, bentonite manufacturers will recommend that protection board be applied over bentonite.

Bentonite sheets are usually nailed in place or fastened using bitumen. Other application requirements should be as recommended by the material manufacturer (see Fig. 4-7). Installation methods vary depending on hydrostatic pressure, substrate materials, edge conditions, and a number of other factors.

Figure 4-6 Bentonite sheets applied against an existing structure. (*Photo by author.*)

Figure 4-7 Bentonite at an existing condition. (*Photo by author.*)

Cementitious Waterproofing

Materials. Here we are interested in two forms of cementitious water-proofing, both of which are nondecorative. The first are plasterlike products consisting of Portland cement, aggregate, and, sometimes, an acrylic or other plastic admix. They are intended for trowel or spray application. These materials are used either on the exterior or the interior of walls. Portland-cement-based cementitious waterproofing materials are available either with or without pulverized iron fillers. When iron is used (metallic oxide waterproofing), an oxidizing agent is included in the mix to make the iron rust quickly and expand to fill the pores in the plaster material.

The second form of cementitious waterproofing we will discuss here is hydraulic cement, which is a compound of cement and rapid-setting non-shrinking hydraulic materials. Hydraulic cements are used for many purposes.

As waterproofing, they are used to seal holes, cracks, and open joints. They will set even when water is actually flowing through the opening being sealed.

Application.   Nondecorative cementitious waterproofing is sometimes used below grade on the exterior face of new exterior walls or on the interior face of new below-grade walls where the exterior face is inaccessible because of shoring or sheeting. Probably the most common use of nondecorative cementitious waterproofing in new work is metallic waterproofing in elevator, sump, and other pits and on the interior surfaces of pipe and other utility tunnels.

Cementitious waterproofing should be installed according to the manufacturer's recommendations. Federal Construction Council's "Federal Construction Guide Specification Section 07-02, Metallic Oxide Waterproofing" and Naval Facilities Engineering Command "Guide Specification Section 07140, Metallic Oxide Waterproofing" are excellent guides for metallic waterproofing installation.

Cementitious waterproofing is brittle and will not span even hairline cracks in the substrate. When the substrate cracks after the waterproofing is in place, the waterproofing will also crack, thereby allowing water to enter the building.

Waterproofing Problems and What to Do about Them

Free water flowing through walls, wet or damp walls, peeling or blistering finishes, and spalling concrete or masonry suggest that the existing waterproofing is no longer preventing water from entering the wall or deck, or that waterproofing does not exist. Condensation; leaks from materials above the top level of the waterproofing or adjacent to the waterproofing; plumbing or mechanical equipment leaks; and other failures may be responsible, of course, and should be investigated and eliminated as causes.

Most waterproofing failures are due to improper substrate or waterproofing system design or installation and often occur because of one of the following reasons:

- The substrate cracks, dislodges, or otherwise fails. When the substrate fails, the substrate must be exposed and completely and properly repaired before waterproofing repair is attempted. Figures 4-8 through 4-10 show several stages of a crack repair.
- Substrates were not properly prepared and cured or required priming was omitted. Applying waterproofing over improperly prepared substrates is a leading cause of leaks and other membrane failures.

Figure 4-8 Routed
hairline crack and
equipment used to
rout it. (*Photo cour-
tesy of Rick Scruggs,
W. R. Grace & Co.*)

Cast-in-place concrete uses only about 25 percent of its water for
hydration. The rest remains in the concrete until it evaporates into the
adjacent air. Unless the concrete is allowed to cure before waterproofing application or a way is provided for that water to escape, liquid-
applied and fully adhered membranes may blister.

Rough surfaces can tear membranes (see Fig. 4-11). Voids and
holes can lead to punctures. Waterproofing will often not bridge over
unfilled cracks. Moving cracks will split waterproofing.

- The wrong waterproofing or flashing material was used. Liquid-applied
membranes used over lightweight insulating concrete may blister, for
example. Materials which are not resistant to weather and sunlight
may deteriorate rapidly when so exposed.

Figure 4-9 Cracks routed, primed, and filled with joint sealant. (*Photo courtesy of Rick Scruggs, W. R. Grace & Co.*)

- The waterproofing was incorrectly installed. Thin liquid-applied membrane, unreinforced joints, improperly sealed laps, and membranes fastened down while stretched are typical examples of installation errors that will lead to failures.
- The wall was coated on both sides with impermeable materials, preventing moisture vapor from leaving the wall. This error will cause crazing or spalling masonry faces and blistering of finishes or waterproofing membranes.
- Asphalt emulsion was not permitted to cure properly. Reemulsified asphalt emulsions may disintegrate or separate from the fabric. Reemulsified asphalts may also bleed through the supporting wall if subjected to hydrostatic head.

Figure 4-10 Crack completely prepared to receive waterproofing. (*Photo courtesy of Rick Scruggs, W. R. Grace & Co.*)

- Insufficient number of plies or material thickness was used to withstand the hydrostatic head present.
- Cants were not installed at the bottom of walls and other direction changes. Cants are necessary for bituminous and modified bituminous materials but may not be necessary for some other membranes. The need should be verified with the membrane manufacturer. Cants are not required for cementitious waterproofing, but coves are necessary at exterior footings.
- Reinforcement was not installed for membrane waterproofing at corners and nonmoving joints in the substrates.
- Building movement from thermal changes, seismic activity, or settlement was not accommodated.

Figure 4-11 Liquid waterproofing applied over this much-too-rough concrete failed and had to be removed. (*Photo courtesy of Rick Scruggs, W. R. Grace & Co.*)

- Waterproofing edges and junctures between waterproofing and other materials were not properly flashed or otherwise treated to exclude water.
- Penetration flashings were not properly designed or installed.
- Waterproofing or flashings were not extended above the water level at adjacent walls or curbs.
- Exterior wall waterproofing was not extended above grade, letting groundwater enter the wall above the waterproofing.
- The waterproofing was installed over an incompatible material.
- The waterproofing system contains incompatible materials.
- Reinforcements or flashings are incompatible with the waterproofing membrane.
- Protection course was not installed, or was improperly installed, permitting the waterproofing to be damaged during subsequent construction operations (see Fig. 4-12).

Figure 4-12 Waterproofing being removed here failed because there was no protection course between membrane and topping. The topping slab was removed before photo was snapped. (*Photo courtesy of Rick Scruggs, W. R. Grace & Co.*)

- Joints between floors and walls were not properly grooved and filled with metallic waterproofing in metallic waterproofing systems. Other joints were not filled.
- Joints between cementitious waterproofing and other materials and penetrations were not filled with sealant.
- Protective plaster or concrete topping was not applied over metallic waterproofing.

Extending and Repairing Waterproofing

Compatibility. Membrane, adhesives, solvents, bitumens, and other materials used to repair, partially replace, extend, or cover existing waterproofing must be compatible with the existing materials in every respect (see Fig. 4-13).

The best way to ensure compatibility is to have the manufacturer of the new material make compatibility tests at the building site. Specifications

Figure 4-13 The dark material is bitumen left after the existing waterproofing was removed. This demonstrates why compatibility is important. (*Photo courtesy of Rick Scruggs, W. R. Grace & Co.*)

should, of course, require that new materials be compatible with substrates and existing materials.

When waterproofing repair or partial replacement is necessary, ascertain the materials originally used and use the same materials. Again, have the waterproofing materials manufacturer make field compatibility tests, even when the same material is being used. Formulations may have changed. The material in place may have undergone unanticipated chemical changes.

Repairing Existing Waterproofing.   Before deciding to try to repair existing waterproofing, get an expert's recommendations. Waterproofing repairs below grade or paving can be very expensive and if improperly done, or not appropriate in the first place, can be ineffective.

A major problem with trying to repair waterproofing is the sometimes impossible task of locating the leak. Even when it is possible to isolate a leak in horizontal waterproofing by flood testing, finding the actual leak requires tearing off the covering material. Often, the process of tear-off will damage the waterproofing, causing other leaks. It is often necessary to remove the covering material and completely replace the waterproofing in the zone where the leak was isolated.

Replacing Existing Waterproofing. Replacing waterproofing can itself be expensive and dangerous. It is difficult to protect the structure below from sudden storms while the membrane is gone. Leaky protection is sometimes better than no protection at all.

When the decision has been made to replace all or a part of existing waterproofing, the covering material, including protection boards (or insulation), must first be removed (see Fig. 4-14). The existing waterproofing must then be scraped away, exposing the structural deck.

Before new waterproofing can be installed, the deck must be prepared. Often, removal of a waterproofing layer will leave the deck with voids, holes, depressions, and rough spots, which must be filled and leveled out for the new waterproofing (see Fig. 4-15). The new waterproofing manufacturer's instructions should be carefully followed. Existing concrete curing

Figure 4-14 Removing bituminous waterproofing using high pressure water and grinding. Residual waterproofing can be seen at left side of photo. (*Photo courtesy of Rick Scruggs, W. R. Grace & Co.*)

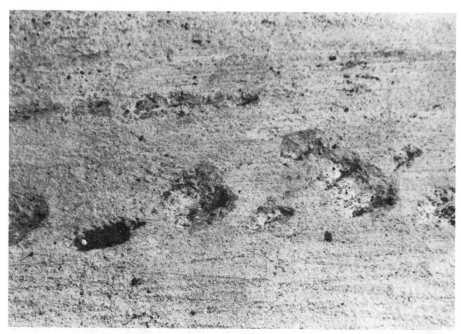

Figure 4-15 Concrete substrate damaged during removal of existing bituminous waterproofing. (*Photo courtesy of Rick Scruggs, W. R. Grace & Co.*)

materials, if any, which may have contributed to the original waterproofing's failure should be removed. The repaired concrete should be primed and otherwise made ready to receive the new waterproofing according to the manufacturer's instructions. The waterproofing should be installed just as it would be over a new substrate. Place the type of reinforcement recommended by the waterproofing manufacturer over small repairs in substrates, and at junctures between new and existing materials. Lap new waterproofing substantially over existing waterproofing or provide a complete separation and protect the joint.

When damage results from coating both sides of a wall with impermeable coatings, the only solution is to remove one of the coatings.

Extending Existing Waterproofing.   When waterproofed walls are extended, extension of the waterproofing will almost certainly be required. When an expansion joint is installed between the original wall and the extension, treatment of the joint must be compatible with both existing and new waterproofing, but the two waterproofing materials do not necessarily have to be compatible. Where the two materials do come into contact, compatability is essential.

New membrane materials should lap existing materials and be sealed to them, unless there is an expansion joint or the membrane manufacturer objects and suggests another acceptable procedure.

Installing New Waterproofing over Existing Materials.   A waterproofing type that can be installed over a particular new material (concrete, for example) can be installed over that same material when it occurs in an existing building. Of course, the substrate should be cleaned and repaired until it is in a condition comparable to that of a new surface of the same material, as acceptable to the new waterproofing manufacturer and installer.

Cementitious waterproofing is sometimes installed on the interior side of existing below-grade walls where leaks are occurring, especially when the walls were not previously waterproofed.

Metallic waterproofing, however, will seldom be installed over an existing floor, because the metallic waterproofing must be protected by a minimum 1-1/2-inch thick concrete topping. Existing equipment, door heights, and flooring may make that impossible. Sometimes, the only way to prevent water from coming through an existing concrete slab-on-grade is to remove the slab and install a drainage system.

Where existing substrates to be waterproofed are coated or painted, it is usually necessary to remove the coating or paint before applying new waterproofing. Coatings and paint should be left in place only when the new waterproofing manufacturer specifically approves.

When the new waterproofing manufacturer concurs, it may be possible to place a slip sheet or barrier between incompatible existing and new waterproofing. Slip sheets and barriers are not, of course, applicable where the new waterproofing must bond to the substrates.

After compatibility has been confirmed and necessary and acceptable barriers and slip sheets have been installed, the new waterproofing should be installed as though the existing material were not there, with all corner and intersection treatment, reinforcement, coats, and thicknesses recommended by the waterproofing material manufacturer for new installations. Where substrate repairs were made, substrate surfaces must be consolidated, properly cured (if the waterproofing material manufacturer recommends it), and smooth before the new waterproofing is installed.

Liquid-applied membranes should not be used on a concrete topping slab that covers a waterproofing membrane, except with specific approval of the new waterproofing material's manufacturer. Such double applications can cause delamination or blistering of the new waterproofing.

Where to Get More Information

The AIA Service Corporation's *Masterspec*, "Basic: Section 07110," "Section 07115," and "Section 07120" address the issues outlined in this chapter related to product selection and use.

The National Concrete Masonry Association's "NCMA—TEK 10A" and "NCMA—TEK 55," though somewhat dated, contain NCMA's official position on interior and exterior below-grade waterproofing of concrete unit masonry walls.

Brent Anderson's 1986 article "Waterproofing and the Design Professional" contains excellent selection and installation criteria for most waterproofing materials and systems. Understanding how waterproofing works can be valuable in discovering what went wrong.

The National Roofing Contractors Association's 1986 *The NRCA Waterproofing Manual* offers good guidance on membrane waterproofing. NRCA was producing an updated and expanded version at the time of this writing, so the current edition may be even more valuable.

CHAPTER 5

Roofing Insulation

Uninsulated roofs have fewer problems than insulated roofs. Condensation is usually eliminated, for example. But most low-slope roofs are insulated. One type of low-slope insulated roofing system is called a "compact" system. In a compact system, roofing membrane, insulation, and vapor retarder, if there is one (see fig. 2-3) are close together, and all are above the deck. In a "framed" system (see fig. 2-2), the insulation is separated from the membrane, often by an air space, usually by the structural deck. Ventilated systems of the type sometimes used where heavy snow occurs are framed membrane systems, because the insulation is separated from the membrane by air space. A system where there are two layers of insulation, one above the deck and another below, is a combination framed and compact system.

Here we are interested primarily in compact systems. Insulation that is separated from the membrane may contribute to roof failure due to condensation but does not affect membrane installation or repair. Refer to Chapter 2 for a discussion of failures due to condensation.

Insulation Types and Their Uses

The components in a compact roofing system may be installed in two different orders. The conventional order (see fig. 2-3) is, starting from the roof deck, insulation, vapor retarder (if there is one), insulation, membrane, and ballast. When the deck is solid (no flutes), the vapor retarder is usually installed directly on the deck.

In a protected membrane roof (PMR) system, the order is deck, underlayment (or insulation), membrane, insulation, polymeric previous fabric, and ballast (see fig. 2-4). The underlayment is needed to bridge the flutes in metal decking and might not be needed over a monolithic deck, such as concrete. With some insulation types, however, underlayment is always needed to achieve a fire-rated assembly. The pervious fabric prevents individual insulation boards from floating or otherwise moving independently. A separate vapor retarder membrane is not needed, because the roofing membrane, being beneath the insulation, serves that function.

Materials

Low-slope roof insulation may be rigid boards, poured-in-place insulating fill, sprayed-in-place plastic foam, or panels that are combination structural deck and insulation, such as cement-wood fiber planks.

Roof insulation should have high impact resistance, thermal stability, low and stable k-value (thermal conductivity), and high strength and rigidity and should not be subject to rapid deterioration in the use intended or be susceptible to damage due to contact with water or water vapor. It should be acceptable to code, fire, and insurance authorities who have jurisdiction over its use.

Board Insulation.   Rigid board roofing insulation types include those manufactured from the following materials:

Vegetable fibers bonded with plastic binders and sometimes impregnated or coated with asphalt.

Mineral fibers bonded with binders and faced with mineral fiber reinforced asphalt impregnated felt.

Glass fibers bonded with phenolic binders and faced with glass fiber reinforced asphalt impregnated felt.

Cellular glass (foamglass) boards and blocks composed of expanded heat-fused glass, which forms closed cells containing hydrogen sulfide.

Perlite, which is a naturally occuring siliceous rock, expanded by heat and combined with organic binders and waterproofing agents, with the top surface treated to minimize bitumen absorption.

Extruded polystyrene (XEPS) foam is a closed cell insulation formed by continuously extruding an expanded polystyrene polymer. It is characterized by its continuous unbroken skin. XEPS is also available with a factory-applied latex modified topping for use in unballasted PMR systems.

Molded polystyrene (MEPS), also called bead board, is also expanded polystyrene polymer but is formed from plastic bubbles, and cut to size and thickness, leaving open cells on the surfaces.

Phenolic board is a closed cell foam that is usually faced on both sides with asphalt impregnated glass or mineral fiber felt.

Polyurethane (PUR) foam is a closed cell board insulation that is seldom used today because it cannot achieve an Underwriters Laboratories Class A fire rating.

Polyisocyanurate foam (PIR) is a closed cell board insulation that is usually faced on both sides with asphalt saturated organic or inorganic felt, aluminum foil, or kraft reinforced aluminum foil facer sheets. Some is reinforced with glass fiber. Polyisocyanurate foam is used in most applications today where polyurethane foam boards would have been used several years ago. Polyisocyanurate is a UL Class A rated material.

Composite boards are factory-laminated multilayer panels formed of a layer of foam insulation such as polyurethane or polyisocyanurate and one or more layers of another insulation such as glass fiber, organic fiberboard, or perlite boards, or another material, such as waferboard, gypsum board, or latex modified concrete. Composite boards are usually faced on the top surface with mineral or glass fiber felt and may be faced on the bottom surface as well.

Most of the above types can be used successfully in compact conventional bituminous built-up membrane systems, but some should not be used there, and some should be used only under some conditions. Polystyrene, for example, is not the best choice for conventional compact built-up bituminous membrane systems, because it does not bond well to decks with hot asphalt and it has a thermal coefficient twice as high as that of the built-up membrane.

Polyurethane board and sprayed-in-place insulations are also questionable in built-up membrane systems, especially in hot humid climates, because under some conditions they are dimensionally unstable. Swollen urethane boards can buckle, and swelling urethane can split membranes covering it. Urethane insulation can be used successfully in membrane roofing systems, but special installation precautions and anchorage are required.

Extruded polystyrene (XEPS), manufactured by the Dow Chemical Company, was the original material used in the patented protected membrane roofing (PMR) system, called Insulated Roof Membrane Assembly (IRMA). XEPS is still the only material suitable for use in PRM but is today manufactured by several companies.

Expanded polystyrene is widely used in ballasted single-ply membrane roofing.

Because of the likelihood that blisters will form when hot asphalt or coal-tar vaporizes water that has been absorbed either by the insulation or the facer sheet, hot asphalt or coal-tar should not be applied directly onto polyurethane, polyisocyanurate, or phenolic foam materials or onto composite boards where those materials form the top layers. The prohibition applies to plain boards and to boards that have facer sheets applied. Those materials can be used with hot bitumen, however, if a layer of mineral fiber, glass fiber, organic fiberboard, or perlite board insulation, or a ventilating base sheet is properly applied over the foam board. A simpler way to avoid the problem, perhaps, is to use a composite board with perlite, fiberboard, or glass fiber as the top layer.

Most modified bitumen and single-ply membrane manufacturers limit the insulation materials they recommend for use with their products. Each membrane material and installation has its own individual set of restrictions. Even the facer sheets used can affect whether or not an insulation board can be used with certain roofing membranes. Aluminum foil faces, for example, are not suitable for built-up roofing applications, and a separator sheet is required between asphalt emulsion-coated fibrous glass mat and polyvinyl chloride (PVC) sheet roofing.

Some insulation manufacturers also advise against using their products with some roofing materials. Some perlite board manufacturers, for example, do not advise using their product with fully adhered EPDM or with any PVC system without an intermediate slip sheet.

Board insulation should not be placed on top of lightweight insulating concrete, poured gypsum roof decks, or thermosetting insulating fill. Additional insulating value, if required in a poured lightweight concrete or gypsum system, should be obtained using the supporting form boards or by surrounding a layer of foam insulation with the poured insulation.

Before any roof insulation product is used in any roofing system, its suitability should be verified with the roofing membrane and insulation manufacturers.

Poured-in-Place Insulation.   Poured-in-place lightweight concrete is concrete made using either perlite or vermiculite as the aggregate.

Concrete made with conventional standard weight aggregate may be made lightweight and used as poured-in-place insulation by adding a foaming agent to the mix.

Thermosetting Insulating Fill. An insulating material gaining acceptance today is thermosetting insulating fill, which is a mixture of perlite and hot steep asphalt.

Sprayed-in-Place Polyurethane Foam. Sprayed-in-place polyurethane foam is included here rather than in Chapter 7 because it is primarily a roof insulation, even though when covered by a waterproof membrane, it is also a complete roofing system.

Plastic films used to protect polyurethane foam insulation from water and the effects of solar radiation, making it a total roofing system, include silicone, acrylic, and urethane. Some single-ply membranes have also been used over polyurethane foam.

Cement-Wood Fiber Planks. Cement-wood fiber planks are structural roof planks which also have thermal insulating qualities and sometimes acoustical properties as well.

Installation

The amount of insulation used depends on three factors: the thermal resistance desired, the R value of the insulation selected, and the fire rating required. Strange as it might seem, a roof system's fire rating sometimes limits the amount of insulation in that system. Too much roof insulation can reduce heat escape so much that fire-generated heat will build up in the ceiling space faster than it would otherwise have and destroy the structure while leaving the roof intact.

Board Insulation. Board insulation in a compact roofing system is either installed conventionally (beneath the roofing membrane) or above the membrane (PMR). Installing the insulation beneath the membrane increases the possibility that condensation will form within the roofing system. Under some conditions (see Chapter 2) a vapor retarder is necessary below, or within, the insulation to prevent condensation from forming. When the insulation is installed above the membrane, the membrane acts as a vapor retarder, reducing the chance that condensation will form within the roofing system.

Insulation installed beneath the membrane, especially insulation with a high R value, retards the membrane's downward dissipation of daytime heat, raising the temperature of the membrane. Some roofing experts think that increasing the membrane temperature contributes to premature aging and early failure of the membrane. Others disagree. It is certain, however, that placing the insulation above the membrane protects the membrane from excessive heat buildup and from impact damage.

There is controversy in the roofing industry about taping board insulation joints. Some manufacturers recommend taping and others discourage it.

Most roofing experts advocate using more than one layer of insulation boards where possible, especially over steel roof decks and in other locations where the insulation is mechanically fastened to the deck. No exception is made for composite boards, even though they are specifically designed for single-layer application. Using multiple layers helps reduce stress in the covering membrane due to insulation joints opening and closing, reduces mechanical fastener length, and insulates the fasteners so that they do not form thermal bridges through the insulation. The lower layer is mechanically fastened in place. The top layer is also sometimes mechanically fastened but more often is bonded to the lower layer by hot bitumen. One disadvantage of double-layer installation is that the intermediate bitumen layer can act as a vapor retarder where one is not needed or is even not desired. Once the vapor retarder is present, the designer must be careful to prevent the dew point from falling below the vapor retarder and must decide whether to introduce vents.

Whether mechanically fastened or not, board insulation on slopes exceeding 1/2 inch per foot should be both mopped and supported by treated wood nailers set level with the top of the insulation.

Anchoring. The circumstances under which board insulation should be positively fastened down is controversial. Board insulation on existing roofs may have been laid loose with no fastening, secured in place using hot asphalt or cold-applied adhesives or mechanically fastened by means of a variety of different fasteners. For years, some roofing experts and manufacturers have advocated laying insulation in loose-laid single-ply roofing systems without fastening. Some still do. Others, however, believe that all board insulation should be fastened down, even when the roofing membrane is not. Still others say that either way is satisfactory. Some insurers and some codes require that insulation be anchored. Factory Mutual (FM), for example, requires mechanical fastening of insulation to steel decks. Some roofing experts believe that FM's requirements are sometimes erroneously followed in roofs where no rating is required—even where fasteners should not be used—in the mistaken belief that if fasteners are good sometimes, they are good all the time. And so the disagreement persists.

Where insulation is anchored, it is fastened to nailable decks and steel decks with mechanical fasteners and to nonnailable decks using bitumen or adhesives.

Insulation that is not anchored is more susceptible to wind uplift damage and may be pulled away from flashings at the roof edges due to cumulative thermal changes in the roofing membrane. Therefore, where the roofing is not anchored, some other device, such as ballast, must hold the roofing and insulation in place.

Some insulation types require special precautions when they are installed using hot mopped bitumen. With polystyrene, for example, it is necessary to allow the bitumen to cool before application.

Recommendations for the number and location of mechanical fasteners, and the type, amount, and placing of bitumen or adhesive are offered by manufacturers, industry associations, such as the National Roofing Contractors Association, Factor Mutual (FM), Underwriters Laboratories (UL), and guide specifications such as *Masterspec* and those produced by the Naval Facilities Engineer Command. Unfortunately, the recommendations do not always agree. Some sources refer to each other for requirements. See ''Where to Get More Information'' at the end of this chapter.

Probably the best way to decide whether anchoring is necessary and to approach anchor design when it is, is to obtain and weigh the recommendations of all available authoritative sources and follow the advice that is the most up to date and makes the most sense under the circumstances. Unfortunately, because things change so fast in the roofing industry, the current consensus may not reflect the most recent data, and published data may not reflect the latest thinking. When there is the slightest doubt, seek knowledgeable professional help. The manufacturer's instructions and warranty terms should be considered, of course, but the insurer and the building code always have the final word.

Underlayment and Priming.   Wood plank, plywood, and cement-wood fiber planks should be covered by a mechanically fastened base sheet before board insulation is applied using adhesives or bitumen.

Concrete substrates should be primed with cutback asphalt before insulation is applied using asphalt bitumen and may also need to be primed before receiving some kinds of adhesives. Priming is not generally necessary when the insulation is applied using coal-tar bitumen.

Polystyrene and polyurethane are combustible. When they are used over a steel deck in a fire-rated assembly, it is necessary to install a layer of class A underlayment, such as gypsum board or rated insulation, to form a barrier between them and the deck. Some codes require a barrier between steel deck and those insulation types, even when a fire rating is not required.

In 1986, one brand of polystyrene insulation board laid directly over steel decking was tested by Underwriters Laboratories in a system where sand was poured into the deck flutes. The system is apparently now approved by UL and may be acceptable by some codes for use in some roofs.

Poured-in-Place Insulating Concrete. Poured-in-place insulating concrete fill is usually placed over steel deck, metal forms, or form boards resting on bulb-tees. Sometimes insulating concrete fill is placed directly over poured-in-place or precast concrete decks, but the practice is not generally recommended because it is difficult to ventilate the insulation in that kind of installation.

Insulating concrete fill is light in weight and can be used to provide a sloping roof surface.

Thermosetting Insulating Fill. Thermosetting insulating fill is applied over a primed solid structural support and hand-tamped or rolled into a compact mass. It can be used to provide a tapered slope for drainage. Metal decking can be used as a substrate, as long as the deck conforms to the fill manufacturer's restrictions. In any deck, openings must be sealed to prevent the hot asphalt from flowing through.

Thermosetting insulating fill can be used in re-covering installations directly over existing built-up roofing.

Sprayed-in-Place Polyurethane Foam. Sprayed-in-place polyurethane roofing has the redundancy of built-up roofing systems since it is field applied in courses but acts as a unit just as does single-ply roofing. It has no joints, which eliminates one of the most failure-prone parts of low-slope roofing systems. In addition, sprayed-in-place polyurethane is fully adhered, lightweight, and self-flashing and can be easily used over irregular, curved, and either new or existing surfaces.

There are, of course, some disadvantages. Fluid-applied membranes protecting the foam must be reapplied periodically. It is difficult to obtain a level surface with the foam. The foam has less impact resistance than do most other roof insulation types. Some roofing experts believe that sprayed-in-place polyurethane roofing systems are a step backward for the roofing industry because their success is totally dependent on the installer's expertise. Almost everyone in the roofing industry agrees that bad installation leads to most roof failures. Modern mechanized application, however, may help to alleviate some potential problems.

Polyurethane foam roofing has been in use since the late 1960s, so it will be encountered on many existing buildings. These systems are most successful, and therefore most used, in low humidity areas such as in the southwest. They are less successful where the climate is hot and humid.

Of particular interest to us here, polyurethane foams can be used successfully for re-covering existing roofing.

Cement-Wood Fiber Planks. Cement-wood fiber planks are installed directly on the structural framing as structural deck units. Planks are anchored in place using clips, nails, or other appropriate devices or are supported by subpurlins anchored to the roof structure.

Roof Insulation Failures and What to Do about Them

Roof insulation failures usually amount to changes in dimension or shape or loss of thermal efficiency. Changes in dimension or shape, which can damage overlying membranes, are usually caused by water or water vapor entering the insulation. Loss of thermal efficiency can happen naturally with age but is often the result of the insulation becoming wet. Many roof insulation failures are due to leaks or condensation. Free water enters through leaks in the roofing's waterproofing membrane or flashing or adjacent construction, while condensation is a roofing system design problem.

Reasons for Failure

Roof insulation fails because of poor roofing system design, improper installation, bad materials, natural deterioration due to age, failure to maintain the roof, and leaks through the roofing membrane, flashings, or adjacent materials.

Poor Design. Refer to Chapter 7 under "Reasons for Failure" for a general discussion of failure in low-slope roofing. The principles discussed there also apply to roofing insulation, which is, after all, merely a part of the roofing system.

Roof slope is particularly important with sprayed-in-place polyurethane foam roofing. Most protective membranes are permeable and will permit water vapor intrusion when free water stands on the membrane for extended periods. The foam will deteriorate when in contact with free water or water vapor. Water vapor behind the protective coating can expand and blister the coating or condense, freeze, and split the coating. Roofs to receive polyurethane foam should be sloped where possible. Some slope in the foam is permissible, but placing foam more than 3 inches thick requires special application procedures, increasing the risk of improper installation. Thick foam, especially when the thickness varies, can create heat buildup during application, which may chemically affect the foam and make it a less effective insulator.

The owner's bad judgment, mentioned in Chapter 7, can extend to arbitrarily demanding too little or too much insulation. Too little insulation will increase fuel consumption and produce high operating costs if (some people say "when") fuel prices, which are now (1988) low, go up. Too much insulation, on the other hand, can impose unnecessary expense. Some roofing experts also believe that increased roof membrane temperatures caused by using too much insulation can lead to more-rapid-than-normal deterioration and premature replacement or re-covering of the existing roof. Not all roofing experts agree, of course. Some say that the insulation's R value has little effect on roof membrane life.

The waterproof membrane and flashing portions of the roofing system, and the adjacent walls and other construction, must prevent water from reaching the roof insulation. Proper design and detailing are critical.

When a vapor retarder (see Chapter 2) is introduced into a compact roofing system, the roofing is sometimes ventilated in an attempt to permit moisture vapor that might enter the system to escape back into the atmosphere without damaging the roofing or insulation. Ventilation is accomplished by grooving the bottom of insulation panels and opening the end of the insulation to the air at the perimeter and by prefabricated vent stacks that connect the insulation space to the air above the roof. Unfortunately, the roofing industry does not speak with one voice about compact roofing systems ventilation. For many years, conventional thinking has been that venting is necessary. Most current guides recommend it. Some recent studies, however, seem to show that not only is venting ineffective, it may actually contribute to roofing problems by permitting water-laden air to enter the system. In addition, vent stacks make small holes in the roofing, which tend to become leak points.

Ventilation, though controversial in board insulation systems, is essential in poured-in-place lightweight insulating concrete. Since venting through them is almost impossible, concrete decks should not be used as a substrate for lightweight concrete insulation. Perforated steel, glass fiber planks, and other permeable decks should be used instead, so that water in the lightweight fill can escape downward. Sometimes, successful ventilation of lightweight fill has been achieved by placing a glass fiber mat or some other ventilating material beneath the fill. Not all lightweight fill manufacturers agree, but some roofing experts believe that ventilating lightweight fills with roof and edge vents is not terribly effective and should be done only in addition to venting the fill from below. Some manufacturers recommend that roof and edge vents and venting from below all be used.

Adequate ceiling plenum ventilation or dehumidification is particularly important when the roof insulation is poured-in-place lightweight insulating concrete, because of the high moisture content of the fill, most of which will dissipate into the ceiling space. Failure to properly ventilate the space below will contribute to condensation and result in damage to ceilings and structural supporting elements.

The binders holding cement-wood fiber planks together will sometimes deteriorate in the presence of excess moisture. They should not be used above wet areas such as swimming pools, showers, or commercial kitchens.

Where additional insulation is placed over cement-wood fiber planks, calculation of dew point location is important to prevent the dew point from falling within the roof system or decking.

Where additional insulation is required with lightweight insulating fill, the additional insulation should be sandwiched between two layers of the fill. Board insulation should not be placed on top of poured insulating fill.

Board insulation should not be installed over thermosetting insulating fill.

Selection of the wrong materials and sizes can sometimes lead to failures. For example, fiberboard insulation, according to ASTM Standard C 208, should be 23 by 47 inches or 24 by 48 inches in size. But many other products and sizes have been improperly used for insulation. Building board, sheathing board, sound-deadening board, and others are not only inappropriate materials for roof insulation, but they are the wrong size (usually 4 by 8 feet). Sizes larger than ASTM Standard C 208 recommends are much more likely to buckle and bow than are the smaller sizes recommended. In roof insulation, bigger is definitely not better.

Improper Installation.   The comments in Chapter 7 about improper installation also apply to roofing insulation.

Improper installation includes using the wrong fasteners in board insulation and the wrong installation methods in any insulation installation. Proper installation technique, while important in every insulation application, is even more important in sprayed-in-place and cast-in-place applications, which are more susceptible to installation errors because they are labor-intensive. In addition, errors are more difficult to correct there and more likely to remain in the permanent installation.

Probably the worst problem in roofing and the most likely to happen is that of trapping moisture within the system during application. Insulation should be laid only during dry weather and should be kept dry continuously until permanently protected by the roofing membrane.

Failure to properly fasten insulation down is also a major problem. Insulation laid in steep asphalt must be laid quickly before the asphalt congeals or the bond will not occur. Mechanical fasteners must be properly spaced and driven to the correct level according to the manufacturer's recommendations.

Placing insulation joints over the flutes in steel decking where they are unsupported can lead to membrane failure due to puncturing or splitting when the insulation deflects into the flute or breaks along the edge of the flute.

Omitting the polymeric pervious fabric in a PMR system can lead to insulation, and consequently roofing, blow-off.

Bad Materials.   Bad insulation materials sometimes arrive at a construction project, but the number of cases is not excessive. Delivery of the wrong materials is more of a problem.

Natural Degeneration and Failure to Maintain.   All materials have finite life spans, but undamaged insulation will usually physically last longer than the roofing material associated with it.

A more serious, and highly controversial, problem with roof insulation is the natural lowering of insulating value when insulation ages. Foams, especially polyurethane and polyisocyanurate are susceptible to a phenomenon known as "thermal drift" whereby the insulation loses R value as it ages. One reason for the loss is replacement of the gas in the cells by air which is not as good an insulator as the original gas. The controversy centers on the method of calculating aged R value. Insulating value lost due to thermal drift can have a large effect on life cycle costs. It can also help determine whether to remove the existing insulation when the roofing has been removed or to leave the insulation in place. To determine the actual aged R value of the in-place insulation it is necessary to test the insulation. If the R value has deteriorated significantly, it might be worthwhile to either add more insulation or remove the existing and install new insulation.

Waterproof Membrane, Flashing, and Adjacent Construction Failure. When the membrane, flashing, or adjacent constructions fails, permitting water to leak into the insulation, there are many negative results.

Insulation materials produce lower R values when they are wet than when dry, because water is a better conductor than insulation, and because the water displaces air—the major insulating factor in many types of insulation. The effect is greater in highly absorptive insulation, such as glass fiber, mineral fiber, and organic fiberboard, and less in most foams. Poured-in-place polyurethane is particularly susceptible to chemical changes due to contact with water during application and curing. Water will not penetrate cured polyurethane foam, but even cured foam will absorb water readily, lessening its insulating capabilities. In addition, polyurethane foam will soften beneath standing water even after curing and expand and pull away from substrates when dampened. Because of its problems with water, polyurethane foam is usually not used in hot humid areas.

In addition, water between the insulation and the roofing membrane can vaporize and cause blisters to form in the membrane. The binders in fibrous insulations may deteriorate in water. Fasteners will rust away. Wood decks will rot. Metal decks will rust. Water in roof insulation or between roof insulation and the roofing membrane can freeze and damage the insulation, the membrane, or both. Some insulation materials will break down completely under such conditions. Foamglass, for example, has been found reduced to powder in severe freeze-thaw conditions. Wet insulation can hold water and permit it to gradually drip into the building for months. Some kinds of insulation can hold enough water to threaten the structure with collapse from overload.

Wet insulation in contact with steel or wood should be removed before it destroys the wood or steel. Some roofing experts believe that when wet insulation exceeds more than 10 percent of the total insulation, the entire insulation should be removed and discarded, but to others that seems excessive. The decision is dependent on many factors, most of which are economic. An often overlooked factor is that repaired roofs do not usually last as long as new ones and are not usually as trouble-free.

Fastener Failure.   Fastener failure is often due to water intruding and rusting the fasteners. Even coating fasteners with noncorrosive materials does not always solve the problem. Driving fasteners through a steel deck will sometimes strip off the coating. Even when the coatings remain intact, water will sometimes travel down the fasteners, or condense on the fasteners, and rust the deck through which they were driven.

Some insulation types generate acids when they decompose due to contact with water. The acids can attack and destroy fasteners.

Fasteners can also back out because they were overdriven initially, stripping the threads (see Fig. 5-1).

Underdriven fasteners will sometimes penetrate the covering membrane or allow the fastened insulation to flutter from wind uplift forces and even tear away from the fastener.

Evidence of Failure

Some roof insulation failures become obvious only when the roofing membrane is removed. Other problems may be found by inspecting or testing.

Inspection and Testing.   When ballast in a PRM, or waterproofing membrane in conventional roofing must be removed, the underlying insulation should be thoroughly inspected for wetting and other damage.

The insulation in undisturbed roofing systems should also be inspected at least once a year to determine if water has leaked into it. Some sources advocate inspecting roofs every 6 months. Inspecting insulation directly, however, is difficult without harming the roofing membrane. Even in protected membrane roofs, thoroughly inspecting the insulation requires removing the ballast. Some insulation damage, though, can be at least guessed at by

Figure 5-1 Improperly installed fastener in the process of backing out. (*Photo courtesy of The Garland Co., Inc.*)

observing roofing surface changes. Ridges, cupping, or curling in the insulation caused by water intrusion will sometimes be visible. Ridges in the membrane over insulation joint lines can be caused by water vapor rising through insulation joints (see Fig. 5-2). When ridges occur along both longitudinal and transverse joints, outlining the insulation boards, the pattern formed is known as "picture framing" (see Fig. 5-3). In cold areas, where frost or a light snow dusting covers the roof, the frost or snow may melt above wet insulation before it melts over the rest of the roof, because water in the insulation lowers its insulating qualities.

Visible water dripping from the roof and water stains on the underside of decking are sure signs that there is water in the insulation. The water may have recently leaked in through adjacent walls, flashings, or the roofing membrane. But it is not unusual for water to enter the insulation through a small leak and build up there for a long time before becoming apparent as stains or drips. Wet insulation often can be detected before drips or stains appear using nondestructive moisture detection methods, such as infrared thermography and nuclear moisture tests. If left alone, water from a small leak can spread horizontally, wetting the insulation over a large area. Early discovery and repair can save extensive work and much money.

Figure 5-2 Moisture through insulation joints causing fully adhered Hypalon to begin to disengage. (*Photo courtesy of The Garland Co., Inc.*)

The same nondestructive methods can also detect condensation build-up in the roofing system so that it can be corrected before major damage is done. Under some conditions, the temperature difference between wet and dry insulation will cause condensation to occur at that point, exacerbating the situation.

When visual inspection or nondestructive testing indicates that water has entered the insulation, test cuts should be made to verify the diagnosis and determine the exact extent of the damage.

When wet insulation is verified, the substrate should also be examined for damage. Water in the insulation may mean rotting wood decking or rusted steel. Sometimes even concrete will be damaged by long-standing water.

Before new roofing is placed over existing insulation, the insulation

Figure 5-3   Advanced "picture framing." (*Photo courtesy of The Garland Co., Inc.*)

should be tested with litmus paper for acidity. Acidic insulation should be removed from contact with metals.

Before new fasteners are driven into an existing deck, core samples should be taken to verify the depth and type of fastener needed and the deck should be tested for pullout. At least ten tests should be made, most of them concentrated in perimeter and corner areas and in areas where leaks or wet insulation were found. More tests should be made if the first tests show low pullout capacity.

Insulation Repair and Reapplication—General Requirements

The term "insulation repair" is almost a misnomer, since existing insulation is seldom actually repaired. Most of the time, damaged insulation is removed and new insulation is installed in its place.

Compatibility.   Insulation, adhesives, bitumen, solvents, and other materials used to install new insulation must be compatible with the existing materials in every respect. The manufacturer or contractor should take samples and have laboratory tests made to verify compatibility. Unless the materials are known absolutely, field examinations or tests will seldom be accurate enough to be considered reliable.

Unfortunately, standards are not available for every roofing product and some available standards are not universally accepted by roofing industry members. So determining compatibility is often difficult. Manufacturers change their products periodically or switch to a different material entirely. The recent switch by most manufacturers from polyurethane to polyisocyanurate foam is an example. There is no central source of roofing materials data. New information and new products appear frequently, making texts and handbooks out of date before they are available for use. The result is that the designer must become much more knowledgeable about roofing and insulation than most of them are comfortable with or run the risk of selecting the wrong materials or wrongly designing the system, with potentially damaging and expensive results to all involved, including the designer.

Code and Insurance Compliance. It is essential that materials used in new insulation applications comply with current fire and insurance requirements, even when the existing materials do not comply. It is also essential that the final in-place roofing system comply with code and insurance requirements. Failure to ascertain and, if necessary, obtain materials and system approvals from authorities having jurisdiction may create liability or cause rejection of the roof by local authorities or denial of insurance. It is not adequate to just use new materials exactly like those in place. The existing materials may no longer comply with the code. After new components have been added, the system may not meet fire-resistance requirements. Adding insulation to a rated system, for example, creates a different system, which may not comply with the fire-resistance requirements of the code or insurer.

Weather Conditions. Because of the danger of trapping moisture in the roofing system, roofing work, including insulation installation, should not be done during inclement weather. Substrates should be permitted to dry completely before new insulation is installed.

Removing Existing Insulation versus Insulating over Existing Insulation. When the existing roofing membrane must be removed, it may be very difficult to salvage the insulation, even when it is dry and otherwise undamaged. Many roofing materials are bonded directly to the insulation and cannot be removed without damaging it.

Where dry undamaged insulation can be salvaged, whether to leave it in place depends on the condition and stability of the existing insulation and its ability to support the loads, including impact loads, associated with roofing replacement. Exposed wet insulation should be removed.

When wet insulation is discovered, and removal of the covering membrane is not otherwise necessary, a decision must be made about whether to repair the leak that caused the insulation to become wet but ignore the wet insulation or to remove the wet insulation and install new insulation before repairing the leak. The decision should be influenced by the possible damage water in the insulation can cause, as discussed earlier in this chapter.

Sometimes wet insulation will be left in place when a new roofing membrane is to be installed, in the mistaken belief that the new membrane will eliminate the problem. Actually, roofing over wet insulation will lead to disastrous consequences. Not only will water left in the insulation cause the kinds of structural damage mentioned earlier, it will quickly invade the new roofing and cause it to fail very rapidly.

Until recently, it was generally accepted wisdom in the roofing industry that wet insulation could be dried out using roof ventilators. Some elaborate gravity ventilators and even some mechanical ventilators were developed for that purpose. Some roofing experts still believe in trying to dry out insulation using vents. But more and more it is becoming apparent that attempts to remove water from existing in-place insulation using gravity or mechanical ventilation systems have usually failed completely or, at best, have been only partially successful.

The inescapable conclusion is that wet insulation should be removed and discarded.

Some roofing experts believe that if more than 10 percent of a roof's insulation is bad, it should all be removed. But removing all the insulation in a conventional compact roof means replacing the entire roofing, which makes little sense when the roofing membrane is essentially sound. Refer to Chapter 7 for further discussion about whether to re-cover or replace roofing.

When existing insulation is no longer adhered to the deck or vapor retarder, it may be necessary to lift and reinstall it even when there is no other damage.

Applying New Insulation over Existing Materials

Before insulation is applied over existing substrates, severely damaged, wet, or acidic materials should be removed and surfaces should be cleaned of dirt and other foreign substances. Proper fastening of left-in-place roofing system components to the roof deck should be verified and corrected if necessary. Existing board insulation which lies over a nailable deck and was not mechanically fastened to the deck should be so fastened before the new insulation is installed, even when the new insulation will be also mechanically fastened in place. Fastener spacings should be the same as those which would be used for installing new insulation on a new deck. Fastener size and length must take the existing roofing thickness into consideration. Existing ballast should be removed, unless the manufacturers of new insulation and roofing membrane specifically agree in writing that removal is not necessary.

When low roof slope is a major contributor to roof failure, it may be advisable to change the roof slope when re-covering or replacing the existing roofing. The slope can be altered using tapered board insulation or thermosetting insulating fill.

When new roofing, with or without new insulation, is installed over an existing roof, calculations must be made to ensure that the design load capacity of the roof is not exceeded.

Sometimes a desire to decrease fuel cost or increase occupant comfort will result in a decision to increase insulation thickness, but adding more insulation can sometimes produce undesirable effects. When new insulation is added to an existing roof, calculations are necessary to make sure that the new condition does not cause condensation to occur or adversely affect the roof's fire rating. Adding insulation above an existing roofing membrane can be particularly troubling, because the existing membrane becomes a vapor retarder.

Insulating over Existing Roofing.   Board insulation may be applied over existing smooth-surfaced, mineral-surfaced, or aggregate-surfaced roofing after proper preparation. Low spots should be filled and ridges and other high spots removed so that the boards will lay flat. The new insulation should be placed in two layers with the long dimension running across the slope and with the end joints staggered. The joints in the two layers should also be staggered from each other.

Over a nonnailable deck, the existing roofing should be primed and both layers of new insulation installed in hot asphalt or adhesive.

Over a nailable deck, the first layer of new insulation should be mechanically fastened to the deck. The second layer may be laid in a hot bitumen or adhesive bed or mechanically fastened. Fastener spacings should be the same as those which would be used for applying new insulation over a new deck. Fastener size and length must take the existing roofing thickness into consideration.

Existing roofing so damaged that it cannot be re-covered using built-up or single-ply membrane systems, or covered with new board insulation, can sometimes be insulated using thermosetting insulating fill or sprayed-in-place polyurethane foam. Preparation for their application requires removing or leveling only the grossest deformations. Surface to be covered with foam must be dry and free of grease, oil, loose particles, dust, and rust. Priming or sealing of most existing surfaces must also be done before foam is applied. Primers and sealers should be those recommended by the foam manufacturer. Asphalt cutback primer should not be used.

The cost of roofing over existing roofing using thermosetting fill or polyurethane foam as insulation is sometimes much less than the cost of removing the existing roofing and installing another roofing system.

Not only is foam very light, but it can be applied over much rougher substrates than board insulation. Even when the structural deck has been damaged to the extent that part of it must be removed, little preparation is needed before the foam is applied to the new decking material.

Thermosetting fill can be roofed using built-up bituminous roofing, modified bitumen roofing of all types, or elastomeric and plastomeric materials, although the latter two may require a slip sheet or separator. Roof membrane installation should be the same as that which would be used for a new roof with thermosetting fill insulation.

Refer to Chapter 7 under "Roofing over Existing Roofing" for a discussion about drilling holes in existing roofing to permit water that gets in to drain out.

Insulating over Existing Insulation.   Low-slope roofing in such poor condition that it cannot be roofed over is often removed, along with its insulation (see Chapter 7). Sometimes, however, it is possible to remove the roofing only, leaving the existing insulation in place.

The principles involved in placing new insulation over existing insulation are the same as those discussed above for installing new insulation over existing roofing.

As was mentioned earlier, lightweight insulating fill and thermosetting insulating fill should not be covered with board insulation.

Insulating over Cleaned Deck.   Placing new insulation over an existing deck from which the original insulation has been completely removed is no different from installing the same material over a new deck.

Where to Get More Information

Refer to Chapter 7, "Where to Get More Information." Many of the references there include data related to roof insulation.

C. W. Griffin's 1982 book *Manual of Built-Up Roof Systems* includes a comprehensive discussion of low-slope roof insulation. It includes excellent discussions of materials, design factors, insulation principles, wind uplift, the contradictory requirements of fire resistance and thermal insulating performance, and nondestructive moisture detection.

The *NRCA/ARMA Manual of Roof Maintenance and Repair* is an excellent guide to inspection, maintenance, and repair of low-slope roofs in general. It includes a discussion of insulation as a component of low-slope roofing systems.

The *NRCA Roofing and Waterproofing Manual* contains data on cement-wood fiber structural panels, lightweight insulating concrete fills, thermosetting insulating fill, and board roof insulation, including general and specific data about cellular glass, composite, glass fiber, perlite, polyisocyanurate foam, polystyrene, and polyurethane foam board insulations. It also includes specifications and diagrams detailing requirements for installing and fastening insulation in many applications, including that over existing materials.

The last word about insulation fasteners and rated roofing systems should come from the local building code, the requirements in Factory Mutual (FM) System's *Approval Guide* and *Loss Prevention Data Sheets,* and the Underwriters Laboratories' (UL) *Building Materials Directory— Class A, B, C: Fire and Wind Related Deck Assemblies* and *Fire Resistance Directory—Time/Temperature Constructions.*

Most good guide specifications also offer some guidance for selecting insulation fasteners, but often they simply refer to FM and UL requirements without specifically stating the requirements.

The Single Ply Roofing Institute's 1987 publication *Single Ply Roofing Systems: Guidelines for Retrofitting Existing Roof Systems* has some advice on dealing with existing wet insulation.

For a detailed discussion of fastener failure, refer to Riaz Hasan's 1987 article "Eliminating Backout and Pullout of Roof Fasteners."

Insulation and roofing membrane manufacturers provide specifications and installation details for their products which should be reviewed and studied. In addition, it is well to seek their advice for specific problems not addressed in their literature. It is not appropriate, however, to use manufacturers' data, or data from other industry sources, for that matter, without comparing that data with recommendations by industry sources and texts written by roofing experts and verifying that the data and other recommendations actually apply to the requirements of the project at hand. Details, of course, must be modified to suit the actual conditions at the project. The roofing industry is replete with conflicting advice and differences of opinion about many subjects, especially about the means, materials, and methods that ought to be used to repair damage. It is a good idea to double-check everything with multiple sources before coming to any conclusion.

CHAPTER

6

Steep Roofing

Steep roofing is common in both residential and commercial structures. This chapter discusses the following steep roofing types:

Roll roofing
Composition shingles
Slate
Wood shingles and shakes
Clay tile
Concrete tile
Mineral-fiber-cement tile
Metal tile

Sheet metal roofing and metal flashings and accessories associated with the steep roofing types addressed in this chapter are covered in Chapter 8.

Although the steep roofing types discussed in this chapter are alike in some aspects, each is covered here separately. Some common data has been included in general paragraphs applicable to all the types covered, but for clarity some data applicable to more than one type has been repeated with each applicable type.

Some steep roofing installation methods that were popular in the past are seldom used today. Discussion of a few of them has been included nevertheless, because an owner, architect, or contractor might have to deal with them when working with an existing building.

Some steep roofing types are used more in some parts of the country than in others. Concrete tile, for example, is popular in southern California, and in Florida, Texas, and other southern states. In an attempt to cover most products used today, some products used in only certain parts of the country are included here. Exclusively local products, however, are not included. Some types, such as reed-thatched roofing, for example, are so rare that even though they may be mentioned, they will not be discussed in detail.

Steep Roof Types and Their Uses

With the possible exception of roll roofing, the steep roof types discussed in this chapter work best when used on roof slopes exceeding 4 inches per foot. Roll roofing may be effective on slopes as low as one inch per foot. Some of the others may be used on slopes as low as 2-1/2 inches per foot when special precautions are taken. Except for roll roofing, materials and systems intended for use for roofs with slopes less than 1-1/2 inches per foot are discussed in Chapter 7.

Requirements Common to All Steep Roof Types

Underlayment.   Roll roofing; composition shingles; slate; and clay, concrete, and metal tile require an underlayment membrane over the roof deck if they are to provide a waterproof installation. On some slopes and in some climates, wood shingles and shakes also require underlayment. Table 6-1 is a synopsis of underlayment requirements.

Except where recommended otherwise in the text associated with each particular steep roofing type in this book, underlayment should be constructed of asphalt-impregnated unperforated roofing felts applied with the long edge parallel with the eaves. Ends should be lapped at least 6 inches. Where continuous underlayment is required, sheets should be applied shingle fashion with joints lapped toward the eaves so that water reaching the membrane will drain. Edge joints in single-layer applications should be lapped at least 2 inches at asphalt roofing, 3 inches at slate roofing, and 2-1/2 inches at

Table 6-1   Underlayment Requirements for New Roofing

Material	Underlayment Requirements				Remarks
	15 lb	*30 lb*	*1 Layer*	*2 Layers*	
Roll roofing	X		X	X	1, 2, 3,
Composition shingles					
4″/ft and steeper	X		X		2
3″/ft to 4″/ft	X			X	3
2-1/2″/ft to 3″/ft	X			X	4
Slate					
4″/ft and steeper					
1/4″ thick and thinner		X	X		2
More than 1/4″ thick		X		X	3, 5
Less than 4″/ft					9
Graduated, all slopes		X		X	3
Wood shingles	X	X			6
Wood shakes	X	X			6, 7
Clay tile	X	X			8
Concrete tile	X	X			8
Mineral–fiber–cement tile					
4″/ft and steeper	X	X			6, 7
3″/ft to 4″/ft	X	X			7
Class A roofs					10
Metal tile	X	X			2

Remarks:
1. Use one layer on slopes 4 inches per foot and more. Use two layers on flatter slopes.
2. Where one layer of underlayment is required, install two layers of 15-pound felt or one 50-pound felt from eave line to 24 inches inside wall line as an ice shield when mean January temperature is 30 degrees Fahrenheit or lower. At slate roofing and metal tile, use two layers of 30-pound felt. Set ice shield felt in hot asphalt or mastic. As an alternative, use adhered bituminous membrane manufactured for the purpose as an ice shield.
3. Where two underlayment layers are required, apply underlayment from eave to 24 inches inside wall using hot asphalt or mastic when mean January temperature is 30 degrees Fahrenheit or less.
4. Entire underlayment applied using hot asphalt or mastic.
5. Alternative: One layer of 55- or 65-pound roll roofing over entire roof, with portion from eave to 24 inches inside wall applied using hot asphalt or mastic when mean January temperature is 30 degrees Fahrenheit or less.
6. Underlayment scheduled is optional except in snow areas (see text).
7. Felt eave strips and felt interlayers between shakes are essential (see text).
8. Underlayment scheduled is often used for roof slopes of 4 inches per foot and higher, but there is disagreement in the industry about underlayment requirements for clay and concrete tile (see text). In applicable areas, the requirements in notes 2 and 3 apply.
9. Slate on slopes less than 4 inches per foot should be installed over membrane roofing or waterproofing (see text).
10. See text.

tile. Double-layer applications should be started with a 19-inch-wide strip cut from a 36-inch roll of felt. Subsequent courses should lap 19 inches over the preceding course, leaving 17 inches exposed. Felts should be nailed under the laps with other felts, using enough nails to hold the felts in place until roofing is applied.

At roll roofing underlayment felts should be placed beneath a metal drip edge at rakes and eaves. At composition shingles, underlayment felts should be installed beneath a metal drip edge at rakes and lap over metal drip edges at eaves by at least 2 inches.

Except at tile roofing, a felt layer should extend at least 12 inches over hips and ridges to provide not less than two thicknesses for each felt layer at those locations. The concealed side of laps should be coated with roofing cement.

Underlayment should extend at least 6 inches up vertical surfaces.

Ice Shield.   Where January mean temperature is 30 degrees Fahrenheit or less, at least two felt layers should be installed beneath steep roofing as an ice shield. Each felt layer in the ice shield should be at least as heavy as that recommended for use beneath the roofing. Ice shield felts should be installed using hot asphalt or mastic from the edge of the roof at the eave line to a line 24 inches inside the exterior wall line below. As an alternative to the multiple layers suggested in the paragraphs discussing each roofing type, a single layer of an adhered membrane intended by its manufacturer to be used for that purpose, or a single layer of 50-pound, or heavier, felt set in hot asphalt or mastic may be used as the ice shield.

Fasteners and Nailers.   All fasteners used in underlayment, drip edges, flashings, and steep roofing should be noncorrosive (see Fig. 6-1).

Wood nailers used with steep roofing should be pressure-preservative-treated. Nailers should also be fire-retardant-treated where the code or fire classification requires.

Asphalt Roll Roofing

Roll Roofing Materials.   Roll roofing suitable for exposed installation consists of an asphalt sheet material coated on the weather side in the area intended to be left exposed with coarse granules of opaque natural colored or ceramic colored slate or other rock. Such roll roofing is called "mineral surfaced roll roofing." Smooth surfaced built-up roofing is discussed in Chapter 7.

Mineral surfaced roll roofing has an expected life span of between 20 and 25 years (see Table 6-2). It is available either as standard type or double coverage type.

Figure 6-1   Failed ungalvanized steel nails in clay roofing tile started this roof damage. Dry rot finished it. (*Photo by author.*)

One entire side of standard mineral surfaced roll roofing is coated with granules. Most standard mineral surfaced roll roofing weighs between 75 and 90 pounds per square (100 square feet).

Only about 17 inches of one face of a 36-inch-wide roll of double coverage roll roofing is coated with granules; the other 19 inches is smooth. Most double coverage mineral surfaced roll roofing weighs between 55 and 70 pounds per square.

Sheathing.   Most asphalt roll roofing is installed over plywood or wood board sheathing, but any continuous nailable substrates will work.

Underlayment.   Surfaces to receive asphalt roll roofing should be covered with underlayment consisting of at least one layer of 15-pound asphalt-impregnated unperforated roofing felt, regardless of the roof slope.

Under the conditions outlined earlier in this chapter under "Requirements Common to All Steep Roof Types," asphalt roll roofing should have an ice shield consisting of at least two layers of 15-pound felt underlayment, regardless of the roof pitch. Where two layers of felt underlayment are

Table 6-2 Installed Roofing Materials Life Spans

Material	Average Life Span in Years	Remarks
Asphalt roll roofing	20 to 25	
Composition shingles	20 to 25	1
Slate	40 to 175	2
Wood shingles	15	
Wood shakes	30	
Clay tile		
Flat	75 or more	
Others	100 or more	3
Concrete tile	50 to 75	
Mineral–fiber–cement tile	30	4
Metal tile	20	5

Remarks:

1. Some shingles are manufacturer warranted for up to 30 years.
2. Average is low (40 to 50 years) for Pennsylvania slate, moderate (100 years) for Vermont and New York slate, and longest for Virginia slate. Slate may require removal and re-laying one or more times to achieve its full useful life.
3. Some manufacturers claim 350-year life spans. Some manufacturers will warrant their clay tile for 50 years. Clay tile may require removal and re-laying to achieve its full life span.
4. This is the standard warranty. This material might last 50 years or more, and some manufacturers warrant their product for that long.
5. If properly maintained, metal tile will last indefinitely. Some manufacturers warrant their metal tile products for up to 35 years.

required over the entire roof because the roof slope is less than 4 inches per foot, the felt from the eave line to a line 24 inches inside the wall line should be adhered using hot asphalt or mastic.

Fasteners. Nails for installing underlayment, drip edges, flashings, and asphalt roll roofing should be 11 or 12 gage, annular- or screw-threaded, galvanized steel or aluminum roofing nails, made for fastening roofing and having at least 3/8-inch-diameter heads and shanks between 7/8 inch and 1-1/4 inches long. Nails for fastening through existing roofing must be of sufficient length to penetrate at least 3/4 inch into the substrate wood.

Asphalt Roll Roofing Installation. Mineral surfaced roll roofing may be applied on any roof slope that exceeds one inch per foot but is usually used on relatively flat roofs where the roofing will not be visible from the ground.

    Roll roofing is placed either parallel or perpendicular to the eave line. When perpendicular, the sheets should be tilted toward the rake at a rate of 1/8 inch per foot of rake.

Following are some additional points to consider when dealing with mineral surfaced roll roofing. For complete installation instructions refer to the National Roofing Contractors Association's 1986 *The NRCA Steep Roofing Manual* and the manufacturers' recommendations.

Standard Mineral Surfaced Roll Roofing.   Standard mineral surfaced roll roofing may be applied using either concealed nails or exposed nails depending on the slope of the roof and the direction of the application. The concealed nail method may be used when the roll roofing material is applied with the long side of the sheet parallel to the eaves, regardless of the roof slope. The concealed nail method may also be used where the roofing material is laid with the long side perpendicular to the eaves, when the slope is 3 inches per foot or more.

In concealed nailing systems, a 12-inch-wide starter strip should be nailed in place completely around the perimeter of the roof and covered with the roofing as it is laid. Roofing cement should be applied over the entire area of each lap of roofing with edge strips and of roofing with adjacent roofing sheets.

The exposed nail method may be used on slopes of 2 inches per foot or more when the material is applied with the long side of the sheet parallel to the eaves. The exposed nail method may also be used when the roll roofing is installed with the long side perpendicular to the eaves, when the slope is 4 inches per foot or more.

In exposed nailing systems, edges should be sealed in place with mastic at eaves and rakes. Laps should be sealed with lap cement before nailing.

Ridges and hip roofing joints should be covered with 12-inch-wide by 36-inch-long strips of roofing. In exposed nail installation systems, the strips are laid with a 2-inch-wide band of roofing cement along each edge of the cover and nailed in place. In concealed nail systems, the strips should be laid in sections in full beds of roofing cement, and each section should be nailed in place beneath the lap of the next section so that the nails are covered by the next section. Laps should be completely bedded in roofing cement.

End laps are usually 6 inches. Edge laps are usually 2 inches.

Double Coverage Mineral Surface Roll Roofing.   Double coverage roll roofing may be used on any roof where the slope is 1 inch per foot or more, providing only that the roof is designed to exclude standing water under all conditions.

Double coverage roll roofing may be installed with the long dimension either parallel with or perpendicular to the eave line. In both cases, all nailing should be done in the selvage area and all laps should be fully bedded in roofing cement. Edges should also be placed over drip edges in a continuous bed of roofing cement. In both installation methods, a starter strip, which is the selvage area from a full sheet, should be first nailed in place. The second sheet should be nailed in place through the selvage area and laid down over the starter strip in a full bed of roofing cement. Succeeding sheets should be laid in the same manner. The exposed portion cut from the starter sheet should be used to complete the installation at the ridge when the sheets are laid parallel with the eave and at the opposite rake when the sheets are perpendicular to the eave.

Flashing.   Flashings at roof drains, vent stacks, and other small penetrations and cap (counter) flashings are usually metal. Base flashings at junctures with walls, chimneys, curbs, and other vertical surfaces are usually bituminous but may be metal. Valley flashings may be either bituminous or metal. Metal flashings are discussed in Chapter 8.

Composition Shingles

Materials.   Composition shingles are the most commonly used steep roofing material. They are available in a variety of sizes, styles, and colors. Shingles are either rag fiber (organic) or glass fiber (inorganic) mats impregnated with asphalt and coated with colored mineral granules.

Individual lock-down-type shingles weighing from 180 to 250 pounds per square are available. Individual lock-down shingles vary from about 18 inches to about 22 inches wide and between 20 and 22 inches long.

Most strip shingles are about 36 inches long by 12 or 12-1/4 inches wide. Some oversize strip shingles are available in widths from 13 to 17 inches and lengths up to 40 inches. Laminated (more than one layer) shingles can be up to 15 inches wide. Single-thickness strips weigh between 215 and 300 pounds per square. Laminated shingles may weigh as much as 390 pounds per square. Heavier weight shingles generally last longer and look better longer than their lighter cousins.

Shingle edge treatment varies. Some of the various types available are shown in Figure 6-2.

Most strip shingles are listed by Underwriters Laboratories as either class A or C. Individual lock-down shingles are also available as either class A or C. Both are available with UL wind-resistance labels.

Composition shingle exposure varies from 4 to 6 inches. Most single-thickness strips, however, are designed for a nominal exposure between 5 and 5-5/8 inches.

Composition shingles with dabs of asphalt adhesive on the surfaces that will be concealed when the shingles are in place are called "self-sealing."

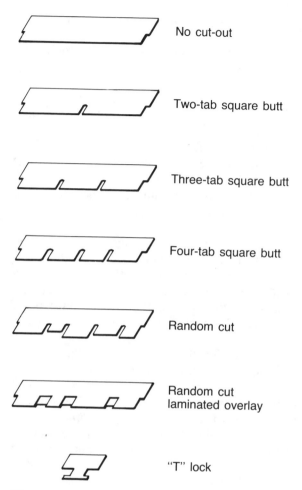

No cut-out

Two-tab square butt

Three-tab square butt

Four-tab square butt

Random cut

Random cut
laminated overlay

"T" lock

Figure 6-2   Shingle edge treatment types.

Sheathing.   Most composition shingles are installed over wood or plywood sheathing but may be installed over any continuous nailable substrate.

Underlayment.   Surfaces to receive composition shingles should be covered with an underlayment consisting of at least one layer of 15-pound asphalt-impregnated unperforated roofing felt, regardless of the roof slope or building location. Surfaces to receive composition shingles should receive a second underlayment layer of 15-pound asphalt-impregnated unperforated roofing felt when the roof slope is less than 4 inches per foot. When the slope is between 2-1/2 and 3 inches per foot, the two underlayment layers should be set in hot asphalt or mastic.

A 36-inch-wide layer of 15-pound asphalt-impregnated felt should also be placed over the previous underlayment and beneath the metal flashing in each valley condition.

Under the conditions outlined earlier in this chapter under "Requirements Common to All Steep Roof Types" composition shingle roofs should have ice shields. The ice shield should consist of at least two layers of 15-pound asphalt-impregnated unperforated roofing felt, regardless of the roof pitch. Where two layers of felt underlayment are required over the entire roof, even though the remaining underlayment is not required to be set in hot asphalt or mastic, the felt from the eave line to a line 24 inches inside the wall line should be adhered using hot asphalt or mastic.

Fasteners.   Nails for installing underlayment, drip edges, flashings, and composition shingles should be 11 or 12 gage, annular- or screw-threaded, galvanized steel or aluminum roofing nails, made for fastening roofing, and having at least 3/8-inch-diameter heads and shanks between 7/8 inch and 1-1/4 inches long. Nails for fastening through existing roofing must be of sufficient length to penetrate at least 3/4 inch into the substrate wood.

Shingle Installation.   Roof slopes restrict the type of shingle that may be used. Any of the shingle types mentioned in this chapter may be used on slopes of 4 inches per foot and more. Some industry sources recommend that shingles not be used on slopes less than 4 inches per foot. The National Roofing Contractors Association's 1986 *The NRCA Steep Roofing Manual*, however, suggests that self-sealing strip shingles with tabs may be used on slopes down to 2-1/2 inches per foot if the additional underlayment mentioned earlier is used.

Strip shingles are usually installed with four nails per shingle. Very steep roofing, such as that on mansards, may require supplementary dabs of asphalt adhesive to hold the shingles in place. Manufacturer's instructions must be carefully followed here, though. Excessive asphalt can damage the shingles.

Shingles can be installed with regular spacing where the first course is started with a whole shingle, the second course is started with a half shingle, and the pattern is repeated throughout the roof. Random patterns are sometimes achieved by starting each course with a different length cut shingle. Some roofs contain ribbon courses at intervals of about five courses. The ribbons are made by inserting three courses of shingles, each cut to a different width, so that three butt edges show.

Flashing.   Flashings at vent stacks and other small penetrations, cap (counter) flashings, ridge flashing, and base flashings at walls, ventilators, skylights, and large penetrations, such as chimneys, are usually metal, which is discussed in Chapter 8. Valley flashings may be either metal or roll roofing.

Slate Shingles

Slate shingles are not among the most widely used roofing materials today, especially on commercial projects, but so much slate roofing exists that some discussion here seems appropriate.

Slate Material.   Slate is hard, practically nonabsorbent, rock characterized by its natural proclivity to split along a single plane. This single-direction cleavage characteristic permits slate to be split into thin, though durable, sheets. Roofing slate is usually split in the direction of the natural grain of the slate.

Roofing slate is classified as either commercial standard, textural, or graduated. Commercial standard roofing slate is smooth and about 3/16-inch thick. Textural roofing slate is 3/8-inch thick or thinner. Its color is more variable than that of commercial standard slate, and its surface texture is rougher. Graduated roofing slate varies in thickness from 3/16 inch to 1-1/2 inches or more. It also varies more in color and texture than either commercial standard or textural slate. Some slate used in roofing (called "ribbon stock") has strips of the rock embedded in it. Slate without ribbons is called "clear slate."

Individual slate roofing pieces (called slates) vary in size from 10 by 6 inches to 24 by 16 inches. Most slates are square. Corners are sometimes clipped to produce special effects. Most slate is split so that the grain runs with the length of the slate.

Slate is available in several colors and shades of colors, including those classified in the industry as black, blue black, gray, mottled gray, purple, variegated purple (mottled purple and green), green, and red. Other colors, called "specials," may also be used. Existing slate will almost certainly have changed color over time and may be difficult to match with new material. The new slate will, of course, also change color, but there may be a time when new slates are apparent. Slate is classified as permanent (or unfading) or weathering, denoting the degree of color fading expected.

When selecting slate to match existing slate it is important to determine the source of the existing slate and try to find the same or similar slate from the same source as a match. Slate is currently mined in Virginia, New York, Maine, Vermont, and Pennsylvania. Existing slate may have come from other locations, also. Slate from one state may not match slate from another state. Even slate from different quarries in the same state, or from a single quarry, may not match. Fortunately, some variation is normal, especially in some colors, so that an acceptable match is often possible.

Every slate should have at least two nail holes machine-punched at the quarry. Slates 3/4 inch or more in thickness when 20 inches or more in length should have four nail holes. Hand punching should be done only where absolutely necessary, such as for fitting hips.

Sheathing.   In the past, slate was often installed using wood pegs hooked over wood lath. More recently, some slate was hung from wood lath or battens or directly from steel structural elements by heavy wire hooked to the wood or structural elements. Sometimes slate was held to steel sections using long bent copper nails. Today, most slate roofing is installed over continuous nailable substrates, often wood, plywood, or a wood fiber product, sometimes over a nailable Portland-cement-based material, and fastened in place using nails.

Underlayment.   Surfaces to receive slate roofing 1/4 inch or less in thickness, where the roof slope is 4 inches per foot or greater, should be covered with an underlayment consisting of at least one layer of 30-pound asphalt-impregnated unperforated roofing felt.

Surfaces to receive slate roofing with a thickness greater than 1/4 inch, where the roof slope is 4 inches or more per foot, should receive an underlayment consisting of two layers of 30-pound asphalt-impregnated unperforated roofing felt, or a single layer of 55- or 65-pound asphalt-impregnated unperforated roofing felt.

Surfaces to receive slate roofing where the roof slope is less than 4 inches per foot cannot be rightly called steep roofs. In such conditions, slate should be installed as tile (without laps) as a ballast over membrane roofing or as a promenade surface over membrane waterproofing.

Under the conditions outlined earlier in this chapter under "Requirements Common to All Steep Roof Types," slate roofs should have ice shields. The ice shield should consist of at least two layers of 30-pound felt.

Fasteners.   Zinc nails, and copper, brass, yellow metal, and other nails with copper content have all been used to install slate with success. For repairing slate work, however, the best choice is large-head copper slaters' nails. Under no circumstances should nails intended for any other purpose, such as roofer's nails or galvanized steel nails, be used with slate roofing.

Nails should be at least one inch longer than the slate thickness but never less than long enough to penetrate the roof sheathing. Nails should not, however, project on the other side of the sheathing, unless they will be concealed in the finished work.

Nail gauge depends on the weight of the slates and locations within the roof. For example, 3 penny nails can be used for commercial standard slates up to 18 inches long. Longer commercial standard slates require 4 penny nails. Hips and ridges of roofs covered with commercial standard slate should be installed using 6 penny nails. Heavier slates require heavier nails.

Slate Installation.   Slates installed with nails should each receive not less than two nails. Four nails should be used in larger slates. Nail size should be as recommended by the slate manufacturer and depend on the slate size and weight. Nails should be driven so that the head just touches the slate. The slate should not be clenched by the nail but should hang on the nail.

Slates should be set so that joints fall as close as possible to the midpoint of the length of the slate immediately above or below, but never less than 3 inches from the end of those adjacent slates (see "Joint Break" in Fig. 6-3). Joints should never be open clear through to the underlayment below.

Slate weather exposure in a roof depends on the size of the slate and the headlap provided (see Fig. 6-3). Exposure is the shingle length minus

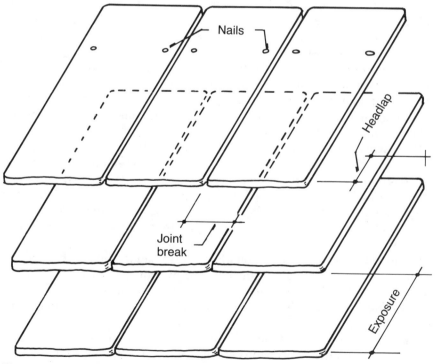

Figure 6-3 Slate jointing.

the headlap divided by 2. The exposure for a 24-inch shingle with a 3-inch headlap is 10-1/2 inches.

The amount of headlap varies with roof slope and region. On slopes from zero to 4 inches per foot, slate should be laid as tile, with no lap. On slopes from 4 inches per foot to 8 inches per foot, headlap is often the "standard" 3 inches, but roofers in some parts of the country consider 4 inches necessary. On slopes from 8 inches per foot to 20 inches per foot the standard 3-inch headlap is generally accepted as sufficient. In some southern areas, a headlap of 2 inches is considered adequate; but over most of the country, 2-inch headlap is restricted to mansards and other roofs with slopes more than 20 inches per foot.

Slate is sold in squares. A square of slate is the amount needed to cover a square of roof (100 square feet of roof surface) when the headlap is 3 inches. The actual headlap to be used must be considered when calculating the amount of slate necessary to cover a given area of roof. More headlap will result in less coverage for a square of slate.

Random-appearing slate roofs using standard and textural slate are achieved by varying slate width while keeping length equal or by varying both widths and lengths. Equal width slates are seldom used.

Variation in graduated slate roofing is achieved by varying lengths, widths, and thicknesses. Larger and thicker slates are used in lower courses and graduate to thinner and smaller slates at the ridge. All courses contain various slate thicknesses.

Repaired hips and ridges should match the existing ones. The National Roofing Contractors Association's 1986 *The NRCA Steep Roofing Manual* and the National Slate Association's *Slate Roofs* each contain detailed descriptions and drawings showing how several common types of hips and saddles should be formed.

Ridges may be either "saddle" or "comb" types. A saddle ridge is one in which the top slates abut evenly. In a regular saddle ridge, the top slates overlap the adjacent top slate (see Fig. 6-4). Two nails, concealed by the next higher slate, are used in each top slate. In a strip saddle ridge, the top slates do not overlap, they abut, and each is fastened down using four nails. Figure 6-5 is a section through either a regular or strip saddle ridge.

Comb ridges are formed similarly to strip saddle ridges, except that the top of the ridge slates do not abut but overlap by between 1/16 inch and 1/8 inch (see Fig. 6-6). The projection may occur all on one side, or the projecting slates may alternate from one side to the other in a configuration called a "coxcomb ridge."

There are several ways to form slate roofing hips. Four common methods are known as "saddle," "Boston," "mitered," and "fantail" hips. Saddle

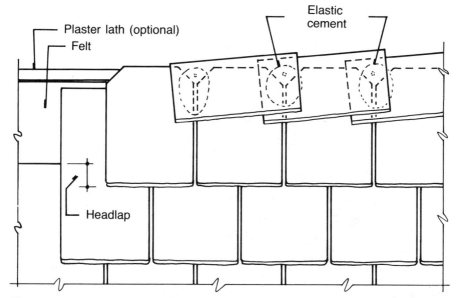

Figure 6-4 Regular saddle ridge.

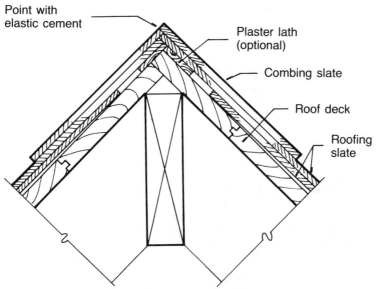

Figure 6-5 Section through a saddle ridge.

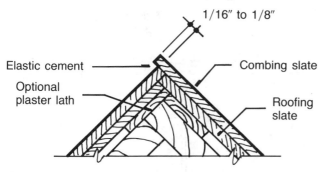

1/16″ to 1/8″

Elastic cement

Optional
plaster lath

Combing slate

Roofing
slate

Figure 6-6 Detail at a comb ridge.

hips and strip saddle hips are similar to saddle and strip saddle ridges, except that each slate is fastened in place using four nails.

In mitered hips, the slates forming the hip are in the same plane as the regular roofing slates and are cut to the slope of the hip (see Fig. 6-7).

Fantail hips are a variation of mitered hips in which the bottom edge of the hip shingle is cut at an angle to form a fantail (see Fig. 6-8).

Boston hips are similar in appearance to saddle hips but are constructed by weaving three shapes of cut shingles together (see Fig. 6-9).

Flashing beneath hips of the kinds mentioned here is usually not necessary. Some existing conditions, however, may have been flashed with metal strips woven in with each course.

Valleys may be either "open," "closed," or "round." Open valleys

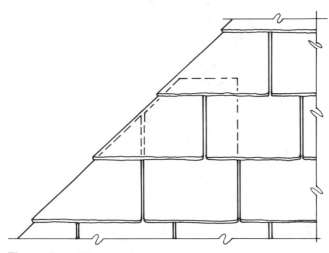

Figure 6-7 Mitered hip.

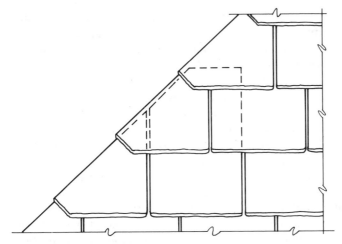

Figure 6-8 Fantail hip.

are projected using metal flashing (see Chapter 8). They should start with the slate 4 inches apart at the top and widen uniformly from top to bottom at a rate of one inch in every 8 feet.

In closed valleys, the slate is cut to abut in the valley and metal flashing is woven in as the slate is installed.

Flashing is not required in properly installed round valleys. Round valleys require considerable underlying support and special slate sizes.

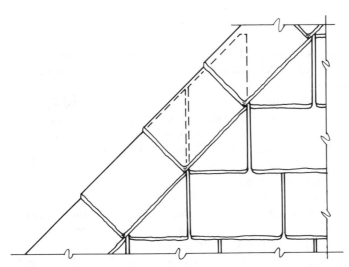

Figure 6-9 Boston hip.

Considerable expertise is needed to properly install round valleys. Round valley slates are sometimes bedded in elastic cement. Round valleys built by inexperienced or insufficiently skilled personnel should be flashed. Round valleys susceptible to ice damage should always be flashed. Round-valley flashing may be made of metal or bituminous materials.

In all types of ridge and hip conditions the top regular slate courses should have their edges set in elastic cement to resist wind uplift. The joint between ridge slates and the overlap in comb ridges should be sealed with elastic cement. Some slate roofing is installed with elastic cement in other locations, such as along flashing lines and in the field of the roofing. Some slate roofing is installed with no elastic cement whatever and experiences no leaks as a result. The absence of elastic cement does not in itself mean that there is a problem or that the installation is inferior.

In every case, however, exposed nails should be coated with elastic cement. Where slates are nailed near flashing, the nails should not penetrate the flashing.

Flashings.   Flashings for use with slate roofing are usually metal. Metal flashings are discussed in Chapter 8.

Wood Shakes and Shingles

Wood shingles and shakes are used throughout the country on residential and commercial properties alike on both sloping and vertical surfaces. Wall applications, however, are beyond the scope of this book. Consult the Red Cedar Shingle and Handsplit Shake Bureau for their recommendations about wall applications.

Wood Shake and Shingle Materials.   Most wood shingles and shakes are made from western red cedar. A small fraction are made from white cedar, redwood, or cypress.

Wood shingles are saw cut on both faces into 16-, 18-, and 24-inch lengths. Butt thicknesses for the three lengths are 0.40, 0.45, and 0.50 inch, respectively. Shingles are graded no. 1, 2, 3, or 4, but only no. 1 shingles, called "blue label," are recommended for roofing. No. 1 shingles are made from 100 percent clear edge-grain quarter-sawn heartwood. The other grades have been used for roofing but are usually relegated to use as siding, where their inherent defects are less noticeable and where leaks are less likely.

Most wood shingles used on roofs are of the square butt type. Shingles with shaped ends (called "fancy butt" shingles) are also available and are used occasionally for roofing (see Fig. 6-10).

Wood shakes are split by hand on at least the exposed surface, which produces a rugged texture. There is only one grade (no. 1) of shakes, but

Figure 6-10 Two examples of "fancy butt" wood shingles. (*Photos by author.*)

there are three types of splits in that grade. "Handsplit and resawn" shakes are split into equal-thickness slabs. The slabs are then sawn diagonally into two tapering shakes. They come in lengths of 15 inches (used for starter strips), 18 inches with 1/2- or 3/4-inch butt thickness, and 24 inches with 3/8-, 1/2, or 3/4-inch-thick butts.

"Tapersplit" shakes are split on both faces by hand into wedge-shaped units. They come 24 inches long with 1/2-inch butt thickness.

"Straight-split" shakes are split by hand also but are not tapered. For use on roofs, they come 3/8 inch thick and either 18 or 24 inches long.

Most shingles and shakes are in random widths, although equal-width shingles are sometimes used.

Hip and ridge covers may be prefabricated units or built up from regular shingles.

Shingles and shakes used on roofs are usually left with natural finish but may be preservative- or fire-retardant treated. Roof shingles or shakes should never be finished with a material that forms a coating.

In the 1920s and 1930s, several manufacturers in this country produced bent wood shingles that were used to make roofs that resembled the curve and flow of traditional reed-thatch roofing. These roofs were usually installed on houses, but some of those houses may be today used as commercial space. Larry Jones's article "American 'Thatch' " in the April 1983 edition of *The Old House Journal* discusses bent wood shingles and says that in 1983 at least, one could obtain further information from *The Old House Journal,* and from C & H Roofing, 1713 South Cliff Ave, Sioux Falls, SD 57105 [(605) 332-5060], and from Great Basin Roofing, 5704 Highland Drive, Holladay UT 84117 [(801) 277-6813].

Sheathing.   Wood shingles and shakes are applied over either open or closed sheathing depending on roof slope and location. Areas where snow occurs should be sheathed with solid sheathing.

Solid sheathing should be at least 1/2-inch-thick plywood.

Spaced sheathing should be 1 × 4s for shingles and 1 × 6s for shakes. Center-to-center spacing of boards should equal the weather exposure of the shingles or shakes but not be more than 10 inches. Three boards should be laid together at the eaves.

Underlayment.   Underlayment requirements for wood shingles and shakes depend on type of sheathing, roof slope, and climate. The Red Cedar Shingle and Handsplit Shake Bureau recommends that all underlayment for wood shingle and shakes should be breather-type building paper, such as deadening felt.

Underlayment may be, but often is not, used beneath wood shingles over either open or closed sheathing. Where pitch is low in an area subject to snow, 30-pound felt underlayment should be used over the entire roof. Where underlayment is used it should comply with the requirements stipulated earlier in this chapter under "Requirements Common to All Steep Roofing Types."

Underlayment is not usually placed over the entire roof sheathing beneath wood shakes. Sometimes, however, a complete underlayment covering of 30-pound felt will be used where the roof is on a relatively low slope and is subject to blowing snow.

Normally, regardless of climate, and regardless of whether there is an underlayment layer over the entire roof, a 36-inch-wide layer of roofing felt is placed along the eave line. After each course of shakes is laid, an 18-inch-wide strip of roofing felt is placed on the shakes and extended over the sheathing. The bottom of the interlayer strip should lie at a distance above the butt of the shakes equal to twice the weather exposure of the shakes (see Fig. 6-11). In temperate climates, 15-pound felt is probably adequate. In cold areas, especially where snow is common, 30-pound felt is often recommended. In warm climates, where straight-split or tapersplit shakes are used, and the weather exposure is less than one-third the shake length, the interlayer is sometimes omitted.

For both shingles and shakes, a layer of roofing felt at least 8 inches

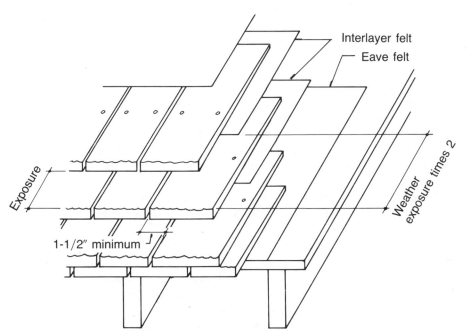

Figure 6-11 Wood shake and mineral-fiber-cement tile installation.

wide should be applied over each ridge and hip. This felt should be of the same weight as that used throughout the roof. Where no other underlayment is used, hip and ridge underlayment can be 15-pound felt.

A 36-inch-wide layer of 15-pound asphalt-impregnated felt should also be placed over the previous underlayment, if any, and beneath the metal flashing in each valley condition.

Fasteners.   Nails should be hot-dipped galvanized or aluminum. It is a foolish economy to use corrodible nails in wood shingle or shake roofs. Such nails will fail long before the shingles themselves need replacing.

The normal nail size for wood shingles is 3d for 16- or 18-inch shingles and 4d for 24-inch shingles.

Shakes are usually installed using 6d nails, which are about 2 inches long, but sometimes longer nails are needed because of shingle thickness or weather exposure.

In every case, nails should penetrate the sheathing a sufficient distance to properly hold the shingle or shake but in no case less than 1/2 inch.

Wood Shingle Installation.   Wood shingles work by shedding water not by blocking its penetration. The water runs off before it can penetrate the roofing. The steeper the slope on which wood shingles are applied, the better is their water-shedding ability.

Shingles are usually applied with all butts in line, but occasionally thatch, serrated, weave, and other patterns are used. Regardless of pattern, the same installation principles apply.

Shingles are installed using only two nails per shingle placed not more than 3/4 inch from the side of the shingle and not more than one inch above the bottom of the next higher shingle course. Nails should be driven flush but should not crush the wood.

Shingles should be doubled at the eaves. Both of the doubled shingles should extend at least 1-1/2 inches beyond the sheathing or fascia board, if applicable. Where a gutter is used, shingle overhang can be reduced to one inch.

Spacing between shingles should be 1/4 inch. Joints should be separated by at least 1-1/2 inches from joints in lower or higher courses. Joints in alternate courses should not align.

Shingle weather exposure depends on shingle size and roof pitch. For roof slopes of 4 inches per foot and greater, standard weather exposure for no. 1 shingles is 5 inches for 16-inch shingles, 5-1/2 inches for 18-inch shingles, and 7-1/2 inches for 24-inch shingles. Most sources do not recommend using shingles on slopes less than 4 inches per foot. Nevertheless, the Red Cedar Shingle and Handsplit Shake Bureau offers recommendations for just such use, saying that on roof slopes from 4 inches per foot to 3 inches per foot, weather exposure for no. 1 shingles should be 3-3/4 inches for 16-inch shingles, 4-1/2 inches for 18-inch shingles, and 5-3/4 inches for 24-inch shingles.

Hips and ridges can be either field-made or factory-prefabricated. Regardless of which is used, hip and ridge shingles should be installed with shingles overlapped to protect nails. Weather exposure should be the same as that of the roofing shingles. Hip and ridge shingles should overlap the shingle on the other side of the hip or ridge in an alternating pattern. Nails must be long enough to properly penetrate sheathing. Usually, 6d nails are sufficient.

Shingles extending into valleys should be cut to follow the line of the valley. Shingle grain should not be parallel to the valley.

Wood Shake Installation.   Like wood shingles, shakes work by shedding water and are not, therefore, recommended for roof slopes less than 4 inches per foot. Occasionally, however, shakes have been used on lower slopes. Such use will probably result in a troublesome installation unless some special precautions are taken. The National Roofing Contractors Association's 1986 *The NRCA Steep Roofing Manual* describes one method for installing wood shakes on low-slope roofs using hot asphalt moppings and wood lattice work or a waterproofing membrane. It is probably a good idea to discuss any special design, including the NRCA-suggested one, with the Red Cedar Shingle and Handsplit Shake Bureau before proceeding.

Shakes are installed using only two nails per shake, placed about one inch from the sides of the shake and one or two inches above the bottom of the next higher course of shakes. Nails should be driven flush but should not crush the wood.

Shakes should be doubled, and may be tripled, at the eaves. The bottom course may be 18-inch shakes or the 15-inch shakes made for that purpose. The shakes should extend at least 1-1/2 inches beyond the sheathing or fascia board, if applicable. Where gutters are used, the overhang can be reduced to one inch.

Spacing between shakes should be 1/2 inch. Joints should be separated by at least 1-1/2 inches from joints in lower or higher courses. Joints in alternate courses should not align.

Shake weather exposure (see Fig. 6-11) depends on shake size and roof pitch. For roof slopes of 4 inches per foot and greater, standard weather exposure for shakes is 7-1/2 inches for 18-inch shakes and 10 inches for 24-inch shakes.

Hips and ridges can be either field-made or factory-prefabricated. Regardless of which is used, hip and ridge shakes should be installed with shakes overlapped to protect nails. Weather exposures should be the same as roofing shakes. Nails must be long enough to properly penetrate sheathing. Usually, two 8d nails on each side of the shake are sufficient.

Shakes extending into valleys should be cut to follow the line of the valley. Shake grain should not parallel the valley.

Flashing.   Wood shingle and shake roofing is usually flashed using metal. Metal flashing is discussed in Chapter 8.

Tile Roofing—General Requirements

Tile roofing comes in many forms, including ancient, traditional, shapes, and newer configurations, and in materials ranging from the stone and clay materials that have been used for centuries to modern-day equivalents made from concrete, mineral-fiber-cement, or metal.

Materials.   Stone and clay tile, along with slate shingles, were early forms of hard roofing materials. Concrete, mineral-fiber-cement, and metal tiles were originally produced and are still used, primarily to mimic clay tile, slate roofing, or wood shingles or shakes. Those newer materials share several general advantages when compared with the earlier forms. They are often lighter, more readily available, and less susceptible to damage. In many cases, the newer materials cost less than the roofing they mimic. Wood shingles are an exception, usually costing less than clay, concrete, or mineral-fiber-cement substitutes, but wood shingles do not have the fire-resistance characteristics shared by the others.

The newer tile forms share the disadvantage of being machine-made. The earlier forms, because they were handmade, produced somewhat irregular surfaces, which appear more rustic and have greater charm than roofs made with their modern mass-produced counterparts. Another disadvantage is that in most cases the modern materials will not stand up to close visual comparison with the products they are imitating and are immediately apparent when installed along side the real thing.

When used alone, though, some of the modern materials are often mistaken for the materials they mimic when viewed from the ground by people who are not experts in recognizing the materials imitated. By contrast, composition shingle that are supposed to look like wood shingles do not fool anybody.

Perhaps the earliest form of heavy roofing, stone tile is generally used today only when repairing historic buildings. Stone tile was never used much in the United States and is not found often enough to warrant detailed discussion here. Refer to "Where to Get More Information" at the end of this chapter for a source of information about stone tile.

Installation.   Unlike those for some other kinds of roofing, industry standards for tile roofing are scarce. Sheathing, underlayments, fasteners, and tile should be installed in accordance with the tile producers' recommendations, applicable building code requirements, and the highest standards of the trade. The requirements in this chapter are a conglomerate of those generally recommended by the producers. Recommendations of industry-recognized sources, such as the National Roofing Contractors Association, have also been taken into account, where they were found. Individual producers and their associations' recommendations and building codes may differ, however. For example, the National Tile Roofing Manufacturers Association (NTRMA) and many tile manufacturers disagree with the recommendations of the National Roofing Contractors Association's 1986 *The NRCA Steep Roofing Manual* to the extent that NTRMA has developed its own recommendations (see "Where to Get More Information" at the end of this chapter).

Flashing.   Flashing for all tile roofing is generally metal. Metal flashings are discussed in Chapter 8.

Clay Tile

Material.   Clay tile is dense, hard, and nonabsorbent material produced by baking moulded clay units in a kiln. It may be either glazed or unglazed.
   Clay tile was used before the Golden Age of Greece. Since at least the early part of the twelfth century, some clay tile shapes have remained virtually the same. Today, clay tile for roofing comes in four general styles: S (Spanish), half-round (Mission), pan and cover (Roman), and flat tile. Figure 6-12 shows some general clay tile shapes, including three in the flat category. Shapes used on an existing building may vary slightly from those shown, especially in the flat and pan types. It is also possible that an existing roof will have a shape not shown.
   Flat shingles may have special shaped butts. These are sometimes called Persian tile. Special butt shapes include those with rounded corners, round ends, half hexagonal ends, half octagonal ends, single pointed ends, and many others.
   Some flat interlocking tile comes in half-, quarter-, and one-and-one-half-wide units to permit finishing at rakes.
   Special shapes are usually used for finishing hips and ridges and for trim and closures at ridges, hips, rakes, eaves, and valleys. Special shapes include regular tiles with the ends closed and specially shaped pieces to fit within or between tiles. Shapes vary to work with each tile type.
   Existing round building elements, such as towers and turrets, may have been roofed using graduated tiles. Graduated tile are units made to match

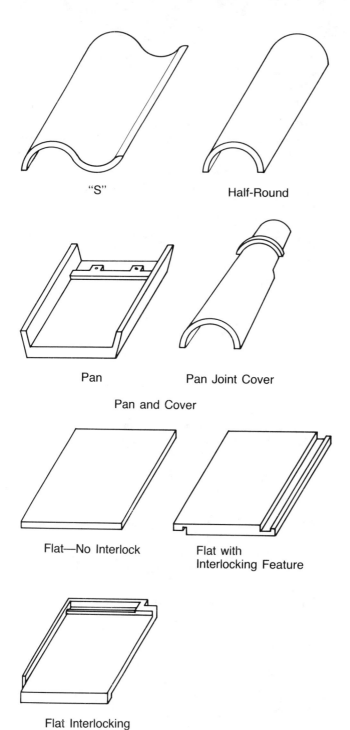

"S"

Half-Round

Pan

Pan Joint Cover

Pan and Cover

Flat—No Interlock

Flat with
Interlocking Feature

Flat Interlocking

Figure 6-12 Clay tile shapes.

the regular types but specially shaped to fit a rounded surface. Tile manufacturers furnish such tile routinely, but matching an existing graduated tile may be a problem. If the particular existing shape is no longer available, it may be necessary to have a match manufactured.

Some clay tile is finished to mimic wood shingles or slate. Broken or cut tiles will show clay's characteristic red color, however, and are easily identified as clay.

Substrates, Battens, and Support Strips.   Clay tile may be installed over wood, metal, or concrete decks. Fastener requirements differ depending on the tile configuration and deck type (see "Fasteners" below).

Wood sheathing for clay tile should be either exterior grade plywood or boards. Plywood thickness should be selected to comply with structural requirements and fastener requirements but should not be less than 1/2 inch. Plywood joints should be left open 1/16 inch to permit expansion.

Board sheathing should be at least 1 × 6s, which should span not more than 24 inches. Supporting structure spans wider than 24 inches require heavier boards. Spaced sheathing is not generally recommended but may have been used on some existing roofs.

Clay tile is seldom installed over cast-in-place concrete decks, because few cast-in-place concrete decks slope enough to receive clay tile roofs. Clay tile applied over cast-in-place concrete is usually fastened in place using wood battens in a configuration known as the "counter batten" system. In a counter batten system, wood battens, usually 1 × 2s, are nailed to wood nailers cast into the concrete deck from eave to ridge at about 20 inches on center. Underlayment is installed over the deck and cast-in nailers before battens are applied. Batten spacing depends on tile size.

Clay tile installed over precast concrete decks (and sometimes over cast-in-place concrete, too) is usually fastened to a counter batten system, in which the wood members running up the slope are through-bolted to the concrete. Sometimes the batten running up the slope is omitted and the horizontal battens are themselves fastened directly to the deck by through bolting.

A system of nailers and wood battens may also be used over metal decks. Often, however, fire ratings or code requirements will dictate that a fire-retardant sheathing be applied over the metal decking. Nailable fire-retardant sheathing, when the sheathing manufacturer so recommends and the tile manufacturer concurs, can be used the same as a plywood deck.

Battens can also be fastened over underlayment to nailers set flush with nonnailable insulation over any deck surface. Insulating products which are also nailable can be used like plywood when their manufacturer so recommends and the tile manufacturer concurs.

Manufacturers of some clay tile types recommend that the tile be applied over battens under all, or sometimes just certain selected, conditions. Such recommendations should be followed. Place such battens at spacings recommended by the tile manufacturer.

A means should be provided for water that finds its way through the tile to drain out of the roofing system. One way is to separate the end joints between battens about 1/2 inch, with end joints about 40 inches on center. The problem with that method is that since joints must fall over the nailers below, the battens must be in many relatively short pieces. A better way is to raise the battens at least 1/4 inch above the underlayment using cut shingles or wood lath strips laid beneath the battens along the line of the nailers and fastened through the underlayment to the nailers.

Tile may also be attached directly to concrete or steel decks using a system of wire hangers (see "Fasteners" below).

S and half-round tile require nailers at ridges, hips, rakes, and projections. Flat tile requires nailers at ridges and hips. Nailer size varies with tile type and rise, which may be different for different manufacturers and is definitely different for different tile types. Nailers are either one inch or 2-inch nominal lumber, depending on tile configuration and tile manufacturer's recommendations. Hip and ridge nailers are usually set in cement mortar.

Eave strips are required except for some types of flat tile where starter course elevation is achieved using a specially made tile strip and other tile where special tile accessories are used to elevate the first course. Eave strips are usually 1 × 2 wood members. Another way to raise the starter course is to elevate the fascia board. In any case, a tapered cant just above the nailer strip or tile starter course, or some other device, is necessary to prevent water from ponding at the eave line.

Underlayment.   All sources agree that surfaces to receive clay tile should be completely covered with an underlayment consisting of at least one layer of 30-pound asphalt-impregnated roofing felt.

The NRCA Steep Roofing Manual says that roof slopes of 3 inches per foot or steeper can receive any type tile when completely covered with an underlayment consisting of two layers of 30-pound asphalt-impregnated felt set in hot asphalt or mastic. The NRCA Steep Roofing Manual also implies that flat clay roofing tile needs a double layer of 30-pound felt underlayment on all slopes less than 5 inches per foot but can be installed on a single layer of felt for steeper slopes.

Tile manufacturers, however, do not all agree with NRCA, or with each other, on those requirements. Some manufacturers agree that any tile shape can be used on any slope exceeding 3 inches per foot when there is a double layer of 30-pound felt as underlayment. Some manufacturers, however, recommend a double layer of 30-pound felt beneath flat tile on all slopes. Some see no need for double layers of felt beneath clay tile except for flat clay tile, regardless of slope. Some say that one layer of 40-pound felt can be substituted for the two 30-pound layers. Some say that underlayment on steep slopes should consist of a single 40-pound felt layer and that underlayment on low slopes should consist of 60-pound felt or two layers of 30-pound felt.

Consequently, existing clay tile may have been installed using any of the above underlayment methods or a method not mentioned here.

There is general agreement that clay tile on slopes below 3 inches per foot should be installed only over built-up roofing or proper membrane waterproofing.

All sources also agree that valleys should have an additional 36-inch-wide layer of 30-pound roofing felt beneath the flashing, and that felts should be carried 6 inches up vertical surfaces.

Where underlayment is used, it should comply with the requirements stipulated earlier in this chapter under "Requirements Common to All Steep Roof Types."

Under the conditions outlined earlier in this chapter under "Requirements Common to All Steep Roof Types," clay tile roofs should have ice shields. The ice shield should consist of at least two layers of 30-pound felt.

Fasteners.   Tile may be nailed in position or hung from wire hangers depending on the tile type, substrates, and code requirements.

Fasteners should be noncorrosive, such as stainless steel, aluminum, or a copper-bearing metal such as hard copper, silicon copper, or yellow metal. Nails should be at least 11 gage, large headed (minimum 5/16-inch diameter), and long enough to penetrate completely through plywood sheathing, not less than 3/4 inch into other wood nailers, and the distance into other nailable materials recommended by the tile manufacturer. Nails should not, however, penetrate completely through board plank decks. Nails for use in plywood should be ring-shank nails. Nails for use in boards should be smooth shank. Nails for use in gypsum plank or nailable concrete should be stainless steel or silicon bronze screw shank nails long enough to penetrate one-half and three-quarters their length, respectively, into deck. They should never penetrate fully through the deck. For very hard decks, it may be necessary to use smooth shank nails.

Wire hangers are sometimes used to hang tile from wood battens but are more often used on concrete and steel substrates. Hangers are also used, even where the substrate is wood, in locations where driven nails would penetrate flashings and where driving nails is difficult. Wire hangers and associated ties are available in stainless steel, brass, copper, and galvanized steel. *The NRCA Steep Roofing Manual* recommends using a 1-1/2-

inch by 1/2-inch wire strip hanger and 14 gage tie wire, but a variety of hanger and tie types are available which may be acceptable to a particular tile manufacturer.

Hurricane clips, also called storm clips, are manufactured from brass or galvanized steel to shapes necessary for use with each tile type. Clips should not be visible in the completed roof. Hurricane clips are fastened positively to the roof deck and clip over tile edges to hold tile in place during high winds.

Mortar and Cement.   Most manufacturers agree that mortar used in clay tile roofing should be composed of a mixture of one part Portland cement, four parts sand, and coloring to produce a mortar that matches the color of the tile.

Plastic cement should be a heavy-bodied asphalt roofing cement. Some tile manufacturers permit using a silicone sealant in lieu of plastic cement. Plastic cement and sealant should be colored to match the tile.

Clay Tile Installation.   The following paragraphs assume that tile will be installed using nails. Similar requirements also apply to tile installed using wire hangers. Wire hanger installation should be done in strict accordance with code requirements and the recommendations of the manufacturers of the tile and the wire hanging devices.

Tile should be installed with one, two, or three nails depending on the type of tile and the tile manufacturer's recommendations. Most field tile will require only one nail. Hip and ridge covers require one or more nails. Tile overlapping flashing should be supported with wire hangers and set in plastic cement. Nail heads should just clear the tile and never be driven down to the tile. Nails must be driven into nailable material, never into joints in the substrates.

The first course of tile should be elevated to proper height by special tile units manufactured for the purpose or by wood strips. Provisions should be made to permit water to drain from the roof beneath the tile. The first tile course should extend beyond the edge of the sheathing or fascia board by 1-1/2 to 3 inches, in accordance with the tile manufacturer's recommendations.

Tile is usually installed with butts parallel to eaves and with tile perpendicular to eaves.

Foreign material should be removed from substrates and from contact surfaces of tile before tile is laid.

Tile should be cut to follow the line of hips and valleys, or special tile may be used. Valleys may be open, closed, or rounded. Most valleys in tile roofs, however, are open. Open ends of tile at valleys should be sealed with special units or mortar, but tile at valleys should not be sealed in a way that would prevent water that has penetrated through the tile to the underlayment surface from draining into the valley. Fan-shaped valley tile should not be sealed at the laps. Battens along valleys should be broken to permit trapped water to escape onto the valley flashing. Tile should lap valley flashing by at least 4 inches.

Open valleys should be left open 6 inches at the top and should taper wider at the rate of one inch for every 8 feet of length.

Roofing tile should be cut and fitted close to and sealed against hip and ridge boards using plastic cement or silicone sealant. Similarly, beads of cement should be placed between tiles at hips and ridges. Hip and ridge covers should be nailed in place and lapped 3 inches. Laps should be sealed with plastic cement. Open space within hip and ridge covers should not, however, be sealed against air circulation. The space between roofing tiles at ridges should be sealed using special units made for the purpose or by mortar. Additional nailers should be applied if special ridge filler tile units are used. Ends of hip and ridge covers should be closed using special units designed for the purpose or by cement mortar.

On flat-tile roofs, one-and-one-half-width tiles rather than half-tile units are usually used at gables. The smaller tiles are too light and tend to crack or blow off in windstorms.

Most tile is set with uniform exposure, but some is set with varying exposure. During tile installation, spacing of tiles may need adjustment to achieve uniform exposure. Exposure is determined by tile size and headlap (the distance a tile laps over the preceding tile course). Recommended headlap may vary from manufacturer to manufacturer, but most recommend that headlap not be less than 3 inches.

Hurricane clips are required on the nose end of each eave-course tile in areas classified by code as "Wind Hazard Areas," and, whether so classified or not, where winds frequently exceed 70 miles per hour. On steep roofs, hurricane clips should also be used periodically through the roof, or the butt of each tile may be embedded in a bead of plastic cement or silicon sealant.

Tile on steep or vertical roofs should be protected against lifting by wind, even when not in "Wind Hazard Areas" by means of wind lock clips.

Other precautions are necessary in high wind areas. For example, tile headlap should be increased to 4 inches, and a bead of mastic should be applied over nail heads at gable, rake, and ridge tiles.

Open-end tile should be sealed at the eaves using special units made for the purpose or sealed with mortar. Either method must permit trapped water to drain out.

Concrete Tile

Material.   Concrete tile is made from a mixture of Portland cement, aggregate, and water and cured under pressure at a controlled temperature and humidity. Some tile is cast; other tile units are extruded. Concrete mixes vary from manufacturer to manufacturer. Some manufacturers use sand as the aggregate. Others use lightweight materials, such as perlite. Some concrete tile is finished on the exposed surface while still wet, using a pigmented cement slurry. Some is colored throughout by pigment added to the concrete mix. Some is the natural concrete color with no added coloration. Concrete tile may be glazed or unglazed.

Some concrete tile is made to mimic wood shingles, clay tile, or slate in shape, color, and texture. Except under close scrutiny, the match with clay tile is fairly successful when the tile is new, differing only in individual tile shape. The attempt to match other materials, however, often fails. For example, some concrete tile is too large to successfully ape wood shingles. Broken edges identify concrete tile made to look like slate or wood. Some concrete tile matches the color of new wood shingles effectively but, of course, does not age as wood does. In addition, the fungus that grows on wet concrete does not look like the fungus that grows on aging wood. Concrete tile does, however, enjoy the major advantage over wood of being fire resistant. And some concrete tile is colored during manufacturing to match closely the appearance of mossy aged wood. A great advantage of using that concrete tile instead of old wood is that the concrete tile will have that same appearance 50 to 75 years later, while wood that far gone will soon need replacing. Such colored concrete tile can be made to look more natural by using different colors placed randomly.

All concrete tile absorbs some moisture. Tile made with some lightweight aggregates absorbs more moisture than does tile made with sand aggregate. The moisture absorption does not seem to harm the tile or its reinforcement but is sometimes unsightly (see Fig. 6-13) and can contribute to rusting metal or rotting wood in the underlying structure. Some concrete tile producers include water-absorption inhibitors in their mixtures to prevent such damage from occurring.

Most concrete tile used for roofing is either the standard concrete type barrel tile shape (see Fig. 6-14) or a flat interlocking shape similar to the unit called "Flat with Interlocking Feature" in Figure 6-12. Shapes used on an existing building may vary slightly from those shown. A common variation, for example, is in the width of the valley in the barrel tile shape. Concrete tile on an existing roof may also be similar to one of the other styles shown in Figure 6-12 for clay tiles or another shape not shown here.

Special shapes are usually used for finishing hips and ridges and for trim and closures at ridges, hips, rakes, eaves, and valleys. Special shapes

Figure 6-13 Water stained concrete mansard tile. Hole in second course was made by an object projecting from a truck. (*Photo by author.*)

include regular tiles with the ends closed and specially shaped pieces to fit within or between tiles. Shapes vary to work with each tile type.

Existing round building elements, such as towers and turrets, may have been roofed using graduated tiles. Graduated tiles are units made to match the regular types but specially shaped to fit a rounded surface. Concrete tile manufacturers furnish such tile routinely, but matching an existing

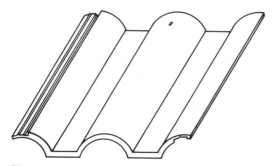

Figure 6-14 Concrete barrel tile shape. Also available are other concrete tile shapes, including some similar to the clay tile shapes shown in Figure 6-12.

graduated tile may be a problem. If the particular existing shape is no longer available, it may be necessary to have a match manufactured.

Substrates, Battens, and Support Strips.   Concrete tile may be installed over wood, metal, or concrete decks. Fastener requirements differ depending on the tile configuration and deck type (see "Fasteners" below).

Wood sheathing for concrete tile should be either exterior grade plywood or boards. Plywood thickness should be selected to comply with structural requirements but should not be less than 1/2 inch. Plywood joints should be left open 1/16 inch to permit expansion.

Board sheathing should be at least 1 × 6s and should span not more than 24 inches. Supporting structure spans wider than 24 inches require heavier boards.

The NRCA Steep Roofing Manual says that spaced sheathing is not recommended in concrete tile installations, but some manufacturers recommend installation over spaced board sheathing and some concrete tile roofs have been installed in that manner. Concrete tile applied over spaced sheathing should be installed in accordance with the tile manufacturer's recommendations.

Some manufacturers and many tile workers would prefer that concrete tile be installed over wood battens on every roof, regardless of slope. A major reason for that recommendation is that using battens makes replacing damaged tile much easier than if battens were not used. Another reason is the general feeling that nails do not hold in plywood well enough to support heavy concrete tile over an extended period. Many low-slope concrete tile roofs have been installed without battens, of course, but battens should be used on slopes of 7 inches per foot and steeper, regardless of tile type or manufacturer's recommendations. Battens are usually 1 × 2s, fastened to the substrate at spacings recommended by the tile manufacturer. Spacing is dependent on tile type and size.

Concrete tile is seldom installed over cast-in-place concrete decks using the nail-on method, because few cast-in-place concrete decks slope as much as 3 inches per foot, which is the minimum at which concrete tile can be nailed to a roof. When concrete tile is applied over cast-in-place concrete by nailing, it is usually fastened to wood battens in a configuration known as the "counter batten" system. In a counter batten system, wood battens, usually 1 × 2s, are nailed to wood nailers cast into the concrete deck from eave to ridge at about 20 inches on center. Underlayment is installed over the deck and cast-in nailers before battens are applied. Batten spacing depends on tile size.

Concrete tile installed over precast concrete decks (and sometimes over cast-in-place concrete, too) is usually fastened to a counter batten system, in which the wood members running up the slope are through-bolted to the concrete. Sometimes the batten running up the slope is omitted and the horizontal battens are themselves fastened directly to the deck by through-bolting.

A system of wood battens may also be used over metal decks. Often, however, fire ratings or code requirements dictate that a fire-retardant sheathing be applied over the metal decking. Nailable fire-retardant sheathing, when the sheathing manufacturer so recommends, and the tile manufacturer concurs, can be used the same as a plywood deck.

Battens can also be fastened over underlayment to nailers set flush with nonnailable insulation over any deck. Insulating products which are also nailable can be used like plywood when their manufacturer so recommends and the tile manufacturer concurs.

A means should be provided for water that finds its way through the tile to drain out of the roofing system. One way is to separate the end joints between battens about 1/2 inch, with end joints about 40 inches on center. The problem with that method is that it requires using many short pieces for battens. The battens must, of course, be nailed to the counter battens. A better way is to raise the battens at least 1/4 inch above the underlayment using cut shingles or wood lath strips laid beneath the battens along the line of the nailers and fastened through the underlayment to the nailers.

Tile may also be attached directly to concrete or steel decks using a system of wire hangers (see ''Fasteners'' below).

Concrete tile requires nailers at ridges and hips. Nailer size varies with tile type and rise, which may be different for different manufacturers and is definitely different for different tile types. Nailers are either one-inch or 2-inch nominal lumber, depending on tile configuration and manufacturer's recommendations. Hip and ridge nailers are sometimes set in cement mortar.

Eave strips are required for concrete tile except for some flat tile types where starter course elevation is achieved using a special tile strip and for other tile types where tile accessories are used to elevate the first course. Eave strips are usually 1 × 2 wood strips. Another way to raise the starter course is to elevate the fascia board. In any case, a tapered cant just above the nailer strip or tile starter course, or some other device, is necessary to prevent water from ponding at the eave line.

Underlayment. *The NRCA Steep Roofing Manual* says that to be waterproof, concrete tile should be installed on slopes of 4 inches per foot or steeper, that flat shingle tile requires a slope of 5 inches per foot or more, and that any slope below 4 inches per foot should have an underlayment consisting of two layers of roofing felt set in hot asphalt or mastic.

Some tile manufacturers, however, do not agree with NRCA on those requirements, nor do they agree with each other. Some say that concrete tile can be nail-applied on slopes of 3 inches per foot or steeper. Some manufacturers recommend that a single-ply 40-pound asphalt-impregnated nonperforated roofing felt underlayment be applied as a minimum. Some recommend that surfaces with slopes between 3 inches per foot and 4 inches per foot to receive nailed-on concrete tile should be completely covered with an underlayment consisting of two layers of 15-pound felt with hot asphalt between the plies, a waterproofing membrane, or a self-adhering bituminous membrane. Some say that slopes of 4 inches per foot or steeper should receive a single layer of 30-pound asphalt-impregnated roofing felt.

Consequently, an existing concrete tile roof may have been installed using any one of those underlayment systems or some other system entirely.

Where an underlayment is used, it should comply with the requirements stipulated in this chapter under "Requirements Common to All Steep Roof Types."

There is general agreement that concrete tile on slopes below 3 inches per foot must be installed only over built-up roofing or proper membrane waterproofing. Some manufacturers say that concrete tile on slopes less than 4 inches per foot should be set in mortar beds over membrane roofing or membrane waterproofing. The method is called "mud-set."

Some manufacturers recommend that "mud-set" tile applications be made over mineral-surfaced roll roofing weighing not less than 83 pounds per square. Others have different requirements.

Under the conditions outlined earlier in this chapter under "Requirements Common to All Steep Roof Types," concrete tile roofs should have ice shields. The ice shield should consist of at least two layers of 30-pound felt.

Fasteners.   Tile may be nailed or wire-tied in position depending on tile type, substrates, and code requirements.

Fasteners should be noncorrosive, such as stainless steel, aluminum, or a copper-bearing metal such as hard copper, silicon copper, or yellow metal. Nails should be at least 11 gage, large-headed (minimum 5/16 inch diameter), and long enough to penetrate completely through plywood sheathing, not less than 3/4 inch into other wood nailers and the distance into other nailable materials recommended by the tile manufacturer. Nails should not, however, penetrate completely through board plank decks. Nails for use in plywood should be ring-shank nails. Nails for use in boards should be smooth shank. Nails for use in gypsum plank or nailable concrete decks should be stainless steel or silicon bronze screw shank nails long enough to penetrate one-half and three-fourths their length, respectively, into the deck. They should never penetrate fully through the deck. For very hard decks, it may be necessary to use smooth shank nails. Some so-called nailable decks may not have sufficient nail holding power to support concrete tile.

Wire hangers are occasionally used to install concrete tile, sometimes to hang tile on wood battens but more often on concrete and steel sub-structures. Even where tile is not generally supported by wire hangers, they are sometimes used to support tiles where nails would penetrate underlying flashing and in locations where nailing is difficult. Wire ties are available in stainless steel, brass, copper, and galvanized steel. *The NRCA Steep Roofing Manual* recommends using a 1-1/2-inch by 1/2-inch wire strip and 14 gage tie wire, but a variety of hanger and tie types are available which may be acceptable to a particular tile manufacturer.

Hurricane clips, also called storm clips, are manufactured from brass or galvanized steel to shapes necessary for use with each tile type. Clips should be designed to be invisible in the completed roof. Hurricane clips are fastened positively to the roof deck and clip over tile edges to hold tile in place during high winds.

Mortar and Cement.   Concrete tile manufacturers do not agree on the mix for mortar for use with concrete tile, or even on the materials to be used. It may be difficult or impossible to determine the actual mortar mix that was used in an existing application. Mortar color usually matches tile color.

In "mud-set" tile applications, concrete tile manufacturers usually recommend using type M mortar as defined in ASTM Standard C 270.

Plastic cement should be a heavy-bodied asphalt roofing cement. Some tile manufacturers prefer that a silicone sealant be used in lieu of plastic cement. Plastic cement and sealant should be colored to match the tile or clear.

Concrete Tile Installation.   The following paragraphs assume that tile, other than that in "mud-set" applications, will be installed using nails. Similar requirements also apply to tile installed using wire hangers. Wire hanger installation should be done in strict accordance with code requirements and the recommendations of the manufacturers of the tile and the wire hanging devices.

In some areas of the country on roof slopes below 4 inches per foot, concrete tile is sometimes installed without nails in mortar beds laid directly over roll roofing. These "mud-set" applications should conform to the tile manufacturer's recommendations.

Nail-applied tile should be hooked over battens (where battens are used) and nailed in place with one, two, or three nails, depending on the type of tile and the tile manufacturer's recommendations. Most field tile will require only one nail. Hip and ridge covers require one or more nails. Tile overlapping flashing should be supported with wire hangers and set in plastic cement. Nail heads should just clear the tile and never be driven down to the tile. Nails must be driven into nailable material, never into joints in the substrates. Some codes require that every tile be nailed. Other codes permit nailing alternate tile rows or even every third tile row. Sometimes, requirements vary, depending on roof slope.

The first course of tile should be elevated to proper height by special tile units manufactured for the purpose, by wood strips, or by a raised fascia. Provisions should be made to permit water to drain from roof beneath tile. The first tile course should extend over the edge of sheathing or fascia board by 1-1/2 to 3 inches according to the tile manufacturer's recommendations.

Tile is usually installed with butts parallel to eaves and with tile perpendicular to eaves.

Foreign material should be removed from substrates and from contact surfaces of tile before tile is laid.

Tile should be cut to follow the line of hips and valleys, or special tile may be used. Valleys may be open, closed, or rounded. Most valleys in tile roofs, however, are open. Open ends of tile at valleys should be sealed with special units or mortar, but tile at valleys should not be sealed in a way that would prevent water that has penetrated the tile to drain into the valley from the underlayment. Fan-shaped valley tile should not be sealed at the laps. Battens along valleys should be spaced to permit trapped water to escape into the valley flashing. Tile should lap valley flashing by at least 4 inches.

Open valleys should be left open 6 inches at the top and taper wider at the rate of one inch for every 8 feet of length.

Roofing tile should be cut and fitted close to and sealed against hip and ridge boards using plastic cement or silicone sealant. Similarly, beads of cement should be placed between tiles at hips and ridges. Hip and ridge covers should be nailed in place and lapped 3 inches. Laps should be sealed with plastic cement. Open space within hip and ridge covers should not, however, be sealed against air circulation. The space between roofing tiles at ridges should be sealed using special units made for the purpose or by mortar. Additional nailers should be applied if special ridge filler tile units are used. Ends of hip and ridge covers should be closed using special units designed for the purpose or by cement mortar. Where cement mortar is used, care should be exercised to prevent water intrusion beneath the tile where it can lead to efflorescence on the face of the tile (see Fig. 6-15).

On flat-tile roofs, one-and-one-half-width tiles are often used at gables in lieu of half-tile units. The smaller tiles are too light and tend to crack or blow off in windstorms.

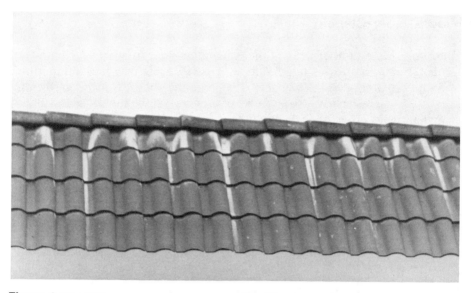

Figure 6-15   Efflorescence on concrete tile. (*Photo by author.*)

Most, but not all, tile is set with uniform exposure. During tile installation, spacing of tiles may need adjustment to achieve uniform exposure. Exposure is determined by tile size and headlap (the distance a tile laps over the preceding tile course). Recommended headlap may vary from manufacturer to manufacturer but most recommend that headlap not be less than 3 inches.

Hurricane clips are required on the nose end of each eave-course tile in areas classified by code as "Wind Hazard Areas," and, whether so classified or not, where winds frequently exceed 70 miles per hour. On steep roofs, hurricane clips should also be used periodically throughout the roof, or the butt of each tile may be embedded in a bead of plastic cement or silicon sealant. Some codes may require additional provisions.

Tile on steep or vertical roofs should be protected against lifting by wind, even when not in "Wind Hazard Areas" by means of wind lock clips.

Other precautions are necessary in high wind areas. For example, tile headlap should be increased to 4 inches, and a bead of mastic should be applied over nail heads at gable, rake, and ridge tiles.

Open-end tile should be sealed at the eaves using special units made for the purpose or sealed with mortar. Either method must permit trapped water to drain out.

Mineral-Fiber-Cement Tile

Material. Mineral-fiber-cement tile is manufactured from an aggregate and cement, reinforced with a mineral fiber. The aggregate may be sand, perlite, or another material. Before asbestos became the scourge of the earth, the fiber used was asbestos. Nowadays, producers use cellulose, glass, polypropylene, and other nonasbestos fibers.

Mineral-fiber-cement tile is manufactured and colored to imitate wood shingles and slate. Colors are either integral or applied. Currently available wood shingle imitations are not convincing, but with some products even experts viewing from the ground cannot be sure whether a roof is covered with slate or the lighter, cheaper mineral-fiber-cement substitute.

Most mineral-fiber-cement tile that is supposed to look like wood is made in 22-inch-long 1/4-inch-thick panels in random widths. Other sizes may be available.

Tiles simulating slate are made in various sizes by their several manufacturers. Some typical sizes are 10-5/8 by 15-3/4 inches, 9-3/8 by 16 inches, 11-13/16 by 23-5/8 inches, and 14 by 30 inches. Thickness is usually about 3/16 inch.

Most mineral-fiber-cement tile is classifed Class A in accord with ASTM Standard E 108.

Mineral-fiber-cement tile weighs from 325 to 560 pounds a square, which is more than composition shingles, about the same as the heavier wood shakes, and less than slate roofing.

Mineral-fiber-cement tile has poor freeze-thaw resistance and should be used in areas subject to such conditions only with the producer's written assurance that such use will not affect the product's warranty.

Sheathing. Mineral-fiber-cement tile may be applied over either open or closed sheathing depending on roof slope and fire rating. Sheathing for Class A roofs is usually solid regardless of roof slope. If mineral-fiber-cement tile is used in areas where snow occurs (see preceding "Material" paragraph for restriction), sheathing should be solid regardless of roof slope. Sheathing for roof slopes of 4 inches per foot and higher may be open-type, except where a Class A roof is required, and except in snow areas, Roofs with slopes less than 4 inches per foot should have solid sheathing.

Solid sheathing should be of the thickness required by the building code and necessary for the supporting framing spans, but in no case should solid sheathing be less than 1/2-inch-thick plywood. Where a Class A roof is required, sheathing must usually be at least 5/8 inch thick.

Spaced sheathing should be 1 × 6s. Center-to-center spacing of boards is usually equal to the weather exposure of the roofing. Boards should be laid tight together in the first 36 inches starting from the eaves.

Underlayment.   Underlayment requirements for mineral-fiber-cement tile depend on type of sheathing, roof slope, and climate.

Underlayment may be used beneath mineral-fiber-cement tile over either open or closed sheathing on any slope but is not often used on slopes of 4 inches per foot or steeper. A 30-pound felt underlayment should be used over the entire roof where the slope is between 3 inches per foot and 4 inches per foot and in areas subject to snow (refer to "Material" paragraph for restriction), regardless of roof slope.

Class A roofs should receive an underlayment over the entire roof of 40-pound asphalt-impregnated glass fiber felt roll roofing.

On slopes less that 3 inches per foot, mineral-fiber-cement tile cannot be relied on to keep water out and should be laid only over a true waterproofing membrane or built-up roofing.

Where underlayment is used, it should comply with the requirements stipulated in this chapter under "Requirements Common to All Steep Roof Types."

Normally, regardless of climate and the presence or absence of an underlayment layer over the entire roof, a 36-inch-wide layer of 30-pound asphalt-impregnated roofing felt is placed along the eave line. After each course of tile is laid, an 18-inch-wide strip of 30-pound asphalt-saturated roofing felt is placed on the tile and extended over, and nailed to, the sheathing. The bottom of the interlayer felt should lie at a distance above the butt of the tile equal to twice the weather exposure of the tile (see Fig. 6-11).

In addition, a layer of 30-pound asphalt-saturated roofing felt at least 8 inches wide should be applied over each ridge and hip.

A layer of 30-pound asphalt-impregnated felt should also be placed beneath the metal flashing in each valley condition.

Fasteners.   Mineral-fiber-cement tile may be fastened in place using corrosion resistant nails or staples as the manufacturer recommends.

Nails and staples should penetrate the sheathing completely, or at least 3/4 inch, whichever is less.

Mineral-Fiber-Cement Tile Installation.   Mineral-fiber-cement tiles are usually applied with all butts in line, but occasionally other patterns might be used.

Tiles are installed using only two nails per tile, placed at least 1/2 inch from the side of the shingle and high enough to be covered at least one inch by the next higher course of tile. Nails should hold tiles firmly but not be overdriven.

Tiles should be doubled at the eaves. Eave tiles should extend at least 1-1/2 inches beyond the sheathing or fascia board, as applicable. Where a gutter is used, the overhang can be reduced to one inch. The starter strip should be installed over a foam strip laid on the fascia board to prevent water penetration beneath the tile.

Spacing between shingles should be 3/8 to 1/2 inch. Joints should be separated by at least 1-1/2 inches from joints in lower or higher courses. Joints in alternate courses should not align.

Shingle weather exposure is usually 10 inches, but the recommendation may vary from manufacturer to manufacturer depending on shingle size and roof pitch.

Hips and ridges can be either field-made or factory-prefabricated. Regardless of which is used, hip and ridge shingles should be installed with shingles overlapped to protect nails.

Shingles extending into valleys should be cut to follow the line of the valley.

Metal Tile

Some metal tile is manufactured primarily to mimic clay tile, slate, or wood shingle or shake roofing. Other metal tile is distinctively shaped and can be mistaken for no other material.

Material. Most metal tile used today is formed from galvanized steel or aluminum sheets into interlocking shapes. In some shapes, metal tile may also be available in solid copper or tern-coated steel. Available shapes include some that appear similar to the standard concrete tile S (Spanish) and half-round (Mission) tile. Metal tile is also available in interlocking strips that simulate wood shingles, square tile laid on the bias, and flat ribbed tile. Other shapes may also be available. Metal tile comes in individual pieces about the size of similar clay or concrete tile and in various-sized sheets and strips, some as long as 16 feet.

Specially shaped pieces are used for drip edges, ridge and hip caps, corner caps, starter strips, end caps, and other accessories. All closers and accessories should be manufactured products of the metal tile manufacturer.

Metal tile is furnished factory-finished with some type of baked-on enamel paint or fluorocarbon coating or with mill finish for field painting.

Sheathing. Metal tile is not structural and must be applied over a solid nailable decking such as plywood sheathing. Sheathing can be 1/2 inch thick for metal tile applied using screws or screw shank nails. Other types of fasteners may require a thicker sheathing. Additional thickness may also be needed to comply with structural requirements, fire requirements, or the building code. Codes or ratings may require a layer of fire-resistive board underlayment, such as gypsum, at Class A rated roofing.

Underlayment. The minimum underlayment for metal tile should be a complete covering of 30-pound asphalt-impregnated roofing felt. An additional 8-inch-wide strip should be applied at hips and ridges, and an additional 36-inch-wide strip should be applied at valleys.

Under the conditions outlined in this chapter under "Requirements Common to All Steep Roof Types," metal tile roofs should have ice shields, which should consist of at least two layers of 30-pound felt.

Fasteners. Fasteners for metal tile should be noncorrosive and should be compatible with the fastener material to prevent electrolytic action from taking place between differing metals. Aluminum fasteners should be used with aluminum tile, for example.

Some manufacturers recommend using screw shank nails with metal tile. Others recommend using only screws. Some even suggest using staples. Certainly, screws make the best installations.

Some manufacturers recommend using pop rivets in certain types of connections.

Metal Tile Installation. Metal tile should be installed in accordance with the manufacturer's recommendations.

Roof slopes for metal tile application should be not less than 3 inches per foot. Where snow occurs, slopes should be at least 4 inches per foot.

Hips and ridges should be made using factory-preformed pieces manufactured for the purpose.

Metal tile must be prevented from contacting dissimilar metals and other materials, such as pressure-preservative-treated wood, masonry, and cement-based products, that might react chemically with the metal. The usual method of preventing contact is to paint the surface of the material that the metal would contact. Sometimes, slip sheets of nonreactive materials will work. The method used should be acceptable to the metal tile manufacturer.

Steep Roofing Failures and What to Do about Them

Steep roofing fails because of bad design, improper installation, poorly manufactured materials, or natural aging.

Reasons for Failure

Bad Design. The earlier portions of this chapter outline steep roofing design and materials requirements and mention sources of additional information. At the end of this chapter, "Where to Get More Information" lists additional data sources. This emphasis on roofing selection and system design reflects their importance in failure prevention.

Bad roof design can result in leaks, excessive heat transfer through the roof, and disintegration of the roofing systems requiring re-covering or replacement of roofing long before its normal life span has been reached. Bad design can also result in condensation damage to the roof or adjacent materials (see Fig. 6-16). Refer to Chapter 2 for a discussion of condensation.

Failing to follow the recommendations of reliable industry-recognized standards or to follow the selected manufacturer's installation details is a major contributor to bad roof design.

Good supporting structure and roofing system design are both critical to successful roofs. The structure must remain stable under all loading conditions and must not flex enough to damage the roofing or permit wind or water intrusion into, and thus through, the roofing. Thermal expansion and contraction must be accounted for. Nailable surfaces must have sufficient nail-holding capacity to support the roofing they will receive.

Bad roofing design includes selecting a poor-quality product; selecting the wrong roofing type for the case at hand; selecting the wrong, or not

Figure 6-16 Condensation in a poorly ventilated attic caused this eave damage. (*Photo by author.*)

enough, insulation; using incompatible, or poorly designed, flashings or roof specialties and accessories; and specifying an improper installation.

Selecting a poor quality product dooms a roof to failure, even when the product is properly installed and regardless of the quality of the workmanship used to install it. Proper investigation of a roofing product, which includes checking references, before use is essential.

Proper selection of steep roofing type and design of steep roofing systems depends on several factors, among which are the climate zone in which the project will be built, the slope of the roof, the appearance desired, and the life span sought. A typical problem, for example, is using a roofing type on a slope that is too low for that roofing type. Another problem is picking a product for an area where its effective life will be shortened due to climatic conditions. Appearance is often a major selecting criteria for steep roofing, but roofing products are not always what they are advertised to be. Composition shingles advertised to simulate wood will not, for example, satisfy a desire for the look of wood shingles.

Insufficient and improperly selected roof insulation will result in excessive heat gain or loss, contribute to condensation problems, and may be a contributing cause to some types of roofing failure. Installing insulation in the wrong configuration within a roof can also lead to roof failure and condensation problems. Roof insulation is discussed in Chapter 5 and condensation in Chapter 2.

Flashings and roof specialties and accessories that are poorly designed or are manufactured using the wrong materials are a major source of roof leaks. It is easy to select the wrong materials for some installations, unless the person making the selection is experienced or thoroughly investigates the problem. A material that works well with one roofing material may cause major problems when used with another roofing type. It is surprising to learn how many professionals are not aware, for example, that you cannot use copper flashing with red cedar shingles because the tannic acid in the shingles destroys the copper. It is also surprising to discover how many professionals design flashings and roof specialties and accessories by "the seat of their pants," never looking at the many sources of good design details that are available to them. Metal flashings and roof specialties and accessories are discussed in Chapter 8.

Although different steep roofing types share some characteristics, installation requirements differ from type to type. Assuming that steep roof types are alike in ways in which they actually differ can be a source of improper design. For example, the underlayment requirements for wood shakes and slate shingles are not alike. And composition shingles, which are light, require four nails in each shingle. Wood shakes, which are much heavier than composition shingles, require not more than two nails. More than two nails in wood shakes will cause excessive splitting. So, specifying too many nails can be as harmful as specifying too few.

Failure to specify ice barriers along eaves in cold zones can lead to ice dams and subsequent roof destruction.

Improper roofing edge design can be a problem too, as demonstrated by Figure 6-17.

Improper Installation. As important as design is, several recent studies have shown that architects and contractors agree that most roofing problems result from poor workmanship and improper installation. Failure to properly seal and protect joints is probably the most common source of problems. The National Roofing Contractors Association advocates continuous visual inspection of low-slope roofing installations as they progress. The same would be helpful for steep roofing installations.

Most steep roofing is installed using some type fasteners. Using corrodible fasteners, such as uncoated or ungalvanized steel nails, is one of the more common problems that show up in steep roofs (see Fig. 6-18). Using corrodible

Figure 6-17 Turning roll roofing over the edge, and omitting a drip, deposited water directly onto the fascia and led to this rot and insect damage. (*Photo by author.*)

Figure 6-18 Wind blew the ridge tile off this Florida train station roof, but rusted nongalvanized steel nails were the real culprit. (*Photo by author.*)

fasteners is also one of the most difficult roofing problems to solve, sometimes requiring complete removal of the affected roofing.

Bad Materials.   Improperly manufactured materials are not unheard of in the roofing industry but occur much less often than do poor design and improper installation.

Inferior products are more common, however. "Inferior" as used here is subjective, of course. The general rule that you get what you pay for is usually, but not always, applicable when purchasing roofing materials. Is a self-adhering ice barrier membrane better than two courses of mopped-in-place asphalt-saturated roofing felt? It costs more, for sure, maybe as much as five times more, but "better" depends on your point of view. On the other hand, heavier composition shingles cost more than lighter ones and are usually superior in terms of life span and maintaining their appearance as they age. Wood shakes cost more than wood shingles. They surely look different, but ascertaining whether one is better than the other depends on who you ask.

Natural Degeneration.   All roofing materials have natural life spans. Table 6-2 lists the average expected life of some steep roofing materials. Damage that occurs after the life span of the material has been exceeded may result from natural degeneration. Then, no amount of remedial work will solve the problem. New roofing is needed.

Evidence of Failure

Most roof failures show up as leaks.

Roof leaks so often result from failed flashing that it makes sense to eliminate flashing failures as a problem before assuming roofing failure. Refer to Chapter 8 for a discussion of metal flashing problems and ways to solve them.

Some roof leaks result from substrate failure, in which case, the underlying structure failure must be located, identified, and repaired before any extensive attempt can be made to repair roof damage. Look for sagging roof lines, a sure sign of underlying structural problems.

Inspection.   Every roof, especially roofs approaching their expected life spans, should be inspected regularly. Patched areas should be thoroughly examined every 6 months. Sometimes, routine inspections disclose impending problems and induce repairs that will prevent leaks from occurring.

When a leak is occurring, do not conduct an inspection. Put a bucket under the drip. The first step in any leak situation is to prevent further damage. If the leak is severe and the damage cannot be otherwise stopped, it may even be necessary to cover the leaking portion of the roof with a tarpaulin or plastic film.

After the leak is under control, inspect the roof to determine the source of the leak. The best time to locate the source of a leak is while the water is coming in, which may mean during a rain. It may be necessary to temporarily remove temporary covers to discover the source of the leak. Some leaks only occur under a single bizarre set of circumstances, such as when the wind is blowing at 37 miles per hour from the north-northwest. Finding the source of that leak under any other conditions may be difficult, if not impossible. Sometimes, leaks can be found by spraying water on the suspected area using a hose.

Even while the water is coming in, finding the particular damage responsible for a roof leak may not be easy. Sometimes, the source of a roof leak in a steep roof will be obvious as curled, cracked, delaminated, loose, or missing roofing materials, but many leaks result from more subtle failures. Some apparent leaks are actually condensation. The potential problem varies with the roofing material, of course, but some aspects of roof leak problems are more or less universal.

It can be difficult to find the source of a roof leak in a steep roof when the source is not obvious immediately. Low-slope roof leaks typically drip near the leak. Steep roof leaks may flow many feet away from the source. The steeper the pitch, the farther the leaked water is likely to travel from the leak source. Sometimes the water will travel completely to the exterior wall and create an ice dam. Sometimes the first sign of water will be at the floor line of a wall where roof water has descended within the walls from the roof far above. Rotting sills two stories below a leak are not uncommon.

Even water on the underside of roof sheathing often enters the roof at a point higher than the apparent leak. Water that penetrates the roofing will sometimes flow above the roof underlayment until it finds a nail hole or other opening in the underlayment and then may flow between the underlayment and the sheathing until it reaches a joint in the sheathing. Nevertheless, examining the underside of sheathing is the place to start looking for the source of roof leaks (see Fig. 6-19). The actual leak will probably be within a foot or so of the apparent leak and will always be higher up the slope.

Figure 6-19 There is obviously a roof leak here, but the display may not help pinpoint the exact source of the leak. (*Photo by author.*)

After the leak source has been located from below, the next step is to examine the roofing itself. The problem may be visible as a badly split wood shake, a broken or slipped slate, a curled asphalt shingle, a shingle lifted out of place or blown away by wind, a missing tile or slate, or split roll roofing.

After finding the source of the immediate problem, examine the entire roof for similar damage. If the damage results in any way from the age of the roof, or is common for the entire roof, there is a good chance that the leak is only the first sign of a more general problem. When the roofing has reached its expected life span, re-covering or replacement of the roofing may be required.

Repair, Replacement, and Re-Covering Work—General Requirements

Compatibility. In repair, replacement, and re-covering work involving bituminous materials first determine the composition of the existing bituminous roofing and roofing cement. Use repair materials that exactly match the existing materials. Do not use coal-tar bitumen where asphalt was originally used, for example, except under conditions acceptable to the product's manufacturer. (Hot coal-tar and hot asphalt are not compatible, but steep asphalt is often used with coal-tar-saturated felt.) Where the exact materials cannot be ascertained, have the manufacturer of the new materials or the roofer who will make the repairs conduct tests at the site to ensure compatibility between the existing and the new materials. When field tests are inconclusive, have samples tested in a laboratory.

Weather Conditions. Do not attempt to repair wet roofing, or do roofing repair work, during inclement weather. Make sure roofing components and substrates are completely dry before beginning repair work.

It is always better, when possible, to make repairs to bituminous materials during mild weather. In hot weather, bituminous materials become soft and easy to tear. Granules on roll roofing and shingles are not locked in as well when the asphalt is soft and are easily knocked off during handling or by workers walking over the roof. In cold weather, bituminous materials become brittle. Cold roof shingles may break when walked on; roll roofing may crack. Tabs may break from cold shingles when they are lifted during installation.

Temporary Fixes. Because leaks are difficult to find and sometimes even harder to repair, there is a tendency to coat suspected leak areas with bituminous roofing cement, in an effort to stop the leak by the shotgun method. Sometimes such a coating will stop a leak temporarily. But using bituminous roofing cement to solve roof leak problems is a bad idea, except in rare cases on bituminous roofs, and then only as a temporary measure.

Whether planned as such or not, bituminous roofing cement coatings are temporary. Unless protected from direct sunlight, bituminous cement will crack and curl and thus allow water to penetrate. When the coating fails, it will expose the original problem to be solved yet again, but then the leak will be even harder to find and fix, because there will be a layer of hard-to-remove bituminous junk in the way. Bituminous cement coatings may actually trap moisture beneath them, creating roof problems worse than what existed before.

In addition, bituminous roofing cement coatings are ugly. Most steep roofs are visible. Black coatings may not be too objectionable on roll roofing, but are horrendous eyesores on shingles, slate, tile, or metal roofing.

Bituminous roofing cement may actually corrode some metals, making its use highly questionable on metal roofing.

Removing Existing Roofing versus Roofing Over Existing Roofing. In every project where new roofing is required, it is necessary to decide whether to remove the existing roofing or to leave the existing roofing in place and apply the new roofing directly over it. The decision rests on the answers to the following questions:

- Is it possible to prepare the existing surface to yield a satisfactory (smooth-enough) surface to receive the new roofing?

    The answer is different for different roofing materials and degrees of damage. The condition of an existing material should be good enough so that it does not interfere with proper installation of the new material. A roofing contractor and the manufacturer of the new product contemplated for reroofing should be consulted at the site and their advice sought before a decision is made to roof over existing roofing or underlayments.

    If existing roofing damage (excessive curling or lifted shingles, for example) has advanced too far, removal may be necessary. If leaks or other conditions have damaged the decking or structural supports, their repair may dictate that roofing and underlayments should be removed.

- What will it cost to prepare the existing roofing to receive the new roofing?

    When the cost of repairing nears the cost of removing, repairing becomes questionable. This factor might be outweighed, however, by the answer to the next question.

- Will leaving the existing roofing in place yield a better completed roof than removing the existing roofing?

    The manufacturers of some roofing products say it does in their case.

- Are the existing roof deck and structure capable of safely supporting the existing roofing, the new roofing, and the other loads associated with installation?

    Roofing over existing materials is advisable only if the substructure is strong enough to support the additional weight, including that of the workers installing the roofing. Even when a structure appears to be sufficiently strong, it is a good idea to have a structural engineer examine the structure to be sure. Some problems may not be apparent to an untrained observer. For instance, a roof might be quite adequate to support the roof dead load and the live loads generally used at the time the roof was designed. But codes and engineering practices change. What was considered adequate 50 years ago, when the building was designed, may violate current law. Other conditions are obvious (see Fig. 6-20).

Figure 6-20 Sometimes the deck must also be removed. This destroyed sheathing was not obvious until the composition shingles and underlayment were removed. (*Photo by author.*)

When the decision is to remove the existing roofing, then it is necessary to decide whether to remove the existing underlayment or to repair the underlayment and lay the new roofing over it. Refer to "Preparing Existing Underlayment to Receive New Roofing" for further discussion.

Preparing Existing Roofing to Receive New Roofing

Where it is feasible to install new roofing without removing the existing roofing, it is necessary to prepare the existing materials to receive the new roofing. Preparation is remarkably similar regardless of the type of new roofing. Where necessary, the following recommendations have been supplemented in the portions of this chapter devoted to each particular roofing type. The recommendations of the affected roofing manufacturers, their associations, and industry-recognized experts such as the National Roofing Contractors Association have been taken into account in developing the following paragraphs, but their recommendations are subject to change from time to time. It would be best, then, to verify the following procedures with current industry wisdom at the time the procedures are implemented.

Materials Requiring Little Preparation.   It is necessary to remove only the grossest irregularities in roll roofing, composition shingles, or mineral-fiber-cement tile before applying new wood shingles or shakes, mineral-fiber-cement tile, or metal tile.

Before new roofing is installed over existing roll roofing, holes should be patched using metal flashing sheets that are compatible with the new roofing, and buckles should be slit and each side nailed down. Either 26 gage galvanized steel or 13-ounce copper sheets are appropriate for metal patches for use beneath most roofing, but copper should not be used where it will contact wood shingles or shakes.

Before new composition shingles, wood shingles or shakes, mineral-fiber-cement tile, metal tile, slate, clay or concrete tile, or any other roofing so permitted are installed directly over existing composition shingles, excessively curled existing shingles and lifted shingles should be removed and new shingles installed in their place.

Locked- or stapled-down shingles may present too rough a surface for satisfactory installation of new composition shingles, slate, wood shingles or shakes, clay tile, concrete tile, mineral-fiber-cement tile, or metal tile. Consult the new roofing manufacturer for advice.

In both roll roofing and composition shingle substrates to receive new roofing, loose nails should be removed and protruding nails should be removed or cut back. New edging strips should be installed where existing edging is corroded or otherwise in poor condition. All flashing should be removed and new flashing installed. To ensure that flashing occurs at the proper level, it may be necessary to build up valleys and other substrates using wood or plywood.

Materials Requiring Major Preparation.   Existing wood shingle or shake roofing to receive new roofing requires considerable preparation. First the existing roof should be carefully examined and gross irregularities such as curled shingles or shakes eliminated. Warped shingles and shakes should be split with a chisel and nailed flat or removed. Voids left by missing or removed shingles or shakes should be filled. Loose shingles or shakes should be securely fastened in place. The first course of shingles or shakes along the eave should be removed and replaced with solid boarding of the same size. The shingles or shakes should be cut back 6 inches from the rakes and replaced with a solid board. Ridge shingles or shakes should be removed and replaced with solid boards. Old valley flashing should be covered with wood strips to separate existing flashings from the new flashings. Walls and penetrations may need work to prepare for new flashings at a different level from existing flashings.

In wood shingle and shake substrates to receive new roofing, loose nails should be removed and protruding nails should be removed or cut back. Flashing should be removed and new flashing installed. To ensure that flashing occurs at the proper level, it will be necessary to build up valleys and other substrates using wood or plywood.

Materials That Must Be Removed.   Existing slate and both clay and concrete tile are probably too heavy to leave in place. In addition, they are hard and brittle, which makes fastening new materials through them almost impossible. Most tile is too irregular in shape to accept new roofing. Usually, it will be necessary to remove slate and tile before installing new roofing.

Existing wood shingles or shakes too rough for conventional preparation must be removed.

Preparing Existing Underlayment to Receive New Roofing

Even where removing existing roofing is necessary, it may be possible to leave existing underlayment in place. Where existing underlayment is damaged but essentially intact, it might be a good idea to apply an additional layer of 15-pound felt over the existing underlayment before applying new roofing. The presence of excessive nail holes and tears from nail removal suggests that an additional layer of underlayment is a good idea. The final numbers of layers and weight of underlayment recommended for new roofing should be provided. Proper ice barriers and metal edges should also be provided where they are appropriate.

Repairing Existing Roll Roofing

Roll roofing failures may occur either in associated flashing, at the joints between roofing sheets, or in the field of the roofing.

Roll roofing and associated bituminous base flashing defects may manifest themselves as alligatored surfaces, holes, cracks, tears, splits, or other damage. Occasionally, blisters occur in roll roofing or associated bituminous flashing.

Repairing Flashings and Drip Edges.   Refer to Chapter 8 for a discussion of metal flashing repair.

Isolated defects in bituminous base flashing may be repaired as follows:

1. Thoroughly clean surfaces to be patched. Remove all loose granules and other loose materials. Allow wet substrates to dry.
2. Check and repair metal cap flashings; bend up to expose entire height of base flashing. See Chapter 8.
3. Prime with asphalt primer and coat the entire surface to receive the patch with a 1/16-inch layer of roofing cement of a type recommended by the sheet materials manufacturer.
4. Bed into the roofing cement a patch of 15-pound roofing felt that covers the entire surface of the base flashing to be patched to a point at least 6 inches beyond the damage and extending at least 4 inches out onto the roof.
5. Cut strips of minimum 55-pound mineral-surfaced roll roofing and coat the backs of the strips with roofing cement. Lay the strips over the previously installed felt patch and press them firmly into place to eliminate air pockets. Extend strips to the top of base flashing, 4 inches beyond the previously installed felt layer at the ends, and 6 inches onto the roof. Lap end joints, if any, not less than 6 inches. Under the counterflashing, nail the strips to the wall with concrete nails through metal discs at about 2 feet on center. Use additional roofing cement to bed roof edges of new flashing patch at roof to ensure positive contact.
6. Similarly patch corners and angles where damage occurs.
7. Bend down counterflashing over base flashing. If no counterflashing occurs, provide it new.

If patches in bituminous base flashing are extensive, it might be better to remove the existing flashing and provide all new flashing.

When damage is due to failure to provide metal drip edges, it may be possible to strip back the roofing at the edge, install a drip edge, and reinstall the roofing.

Repairing Roofing and Underlayments.   Some failures in roll roofing show up as opened joints where the sheets overlap. Sometimes it is possible to force flashing cement under the roofing and roll or press it flat. When that does not work, the seam must be slit and nailed on each side of the slit. Then the patch must be covered with a layer of mineral surface roll roofing overlapping the new nail heads by 6 inches in each direction. The entire back side of the strip should be coated with bituminous plastic cement and pressed into place. Then the strip should be nailed in place with roofing nails. Then the nail heads should be coated with bituminous plastic cement.

Small defects in roll roofing may be patched as follows:

1. Cut out the damaged area. Do not cut through underlayment.
2. Clean the cut area and remove all loose particles and dirt. Allow area beneath and around the cut-out to dry if wet or damp. Force roofing cement under the edges of the cut and coat the entire area of the cut-out.
3. Cut a patch from roll roofing to exactly fit the cut-out area. Bed the patch firmly into the roofing cement.
4. Cut a covering patch from roll roofing, sized to overlap the cut-out by 2 inches in each direction. Coat the back side of the covering patch with a heavy layer of roofing cement and press it firmly into place over the cut-out patch. Nail the covering patch in place with roofing nails at 2 inches on center all around the edges. Cover the nail heads with roofing cement and sprinkle on a coating of fine gravel, stone granules, or sand.

When damage results from failure to provide an ice shield, it might be possible to strip back the roofing at the edge, install an ice shield, and reinstall the roofing.

Nail Problems.   A common problem in roll roofing failure is that of rusting of ungalvanized steel nails used to install metal drip edges or even the roofing itself. Over time, usually before the life expectancy of the roofing has expired, ungalvanized steel nails will rust through and no longer hold the roofing in place. Renailing is possible under such circumstances but may require a covering layer of felt over the new nails to provide a watertight roof. Often, by the time improper fasteners become apparent, the roof has been damaged so extensively that renailing is not a practicable solution. Such discovery often comes when portions of the roof are blown off in a wind storm.

Even galvanized nails will not last forever, but damage due to their rusting may be a suggestion that the roofing has reached its expected life span and should be replaced or re-covered.

Installing New Roll Roofing over Existing Roofing

There may be no solution to some design failures short of installing new roofing. If underlayment has been omitted, for example, even small damage to the surface will probably result in a leak. In such an event, if the structure is capable of holding the load, it might be possible to use the existing roofing as an underlayment and install a new roof on top of it. Otherwise it will be necessary to remove the existing roofing and provide new roofing.

When defects are extensive, or affect all areas of the roof, it is probably time to install new roofing. Installing new roofing may also be best when failures occur after roll roofing has reached its natural 20- or 25-year life span.

There are two options for installing new roll roofing where roll roofing exists: remove the existing materials or leave the existing materials in place and roof over them. Roll roofing is seldom installed where other types of roofing exist.

Removing All Existing Materials.   When the first option is selected, existing roofing and underlayments are completely removed and the deck is inspected and repaired, if necessary, and prepared as a new deck would be.

Roofing over Existing Materials.   Roofing and underlayments may be left in place and the new roofing applied over them, or only the roofing may be removed and the underlayment left in place.

For recommendations for preparing existing roofing to receive new roll roofing, refer to "Preparing Existing Roofing to Receive New Roofing" earlier in this chapter.

For recommendations for preparing existing underlayment to receive new roll roofing, refer to "Preparing Existing Underlayment to Receive New Roofing" earlier in this chapter.

Roll roofing over existing underlayment or existing roll roofing should be applied in the same way that new roofing would be installed on new substrates.

Repairing Composition Shingle Roofing

Composition shingle roofing failures may occur either in associated flashing or in the shingles themselves.

Repairing Flashings and Drip Edges.   Refer to Chapter 8 for a discussion of metal flashing repair.

Bituminous flashing damage may show as alligatored surfaces, holes, cracks, tears, splits, or some other damage. Occasionally, blisters occur in bituminous flashing. Damaged bituminous base flashings may be repaired as recommended in this chapter for repairing defects in the field of roll roofing, but repair is probably not advisable. Most of the time, it would be better to remove the damage flashing and install new flashing. Use the same methods for installing new flashing as are appropriate for installing similar flashing on an all-new roof.

Repairing Shingles and Underlayment. There are two basic kinds of shingle damage: non-age-related damage and age-related deterioration. Non-age-related damage often occurs in roofs that are approaching their expected life span, because the shingles are more susceptible to damage as they get older.

Non-age-related damage may also affect relatively young roofs, as a result of severe windstorms, hail, earthquakes, broken tree limbs, or people walking or working on the roof. Some types of non-age-related damage may affect large portions of the roofing and require that new roofing be installed. Much non-age-related damage, however, affects single shingles or small groups of shingles.

Non-age-related damage may include splits, cracks, holes, chips, missing surface granules, or deformed shingles. Shingles may blister due to excessive use of roofing cement during installation.

Windstorm damage may occur during winds exceeding the rated wind resistance of the shingles, which in some shingles is only 54 miles an hour.

Shingles may also be damaged by workers repairing roof-mounted equipment, or even during inspections of the roof itself.

It may be possible to remove individual, or groups of, shingles damaged in non-age related incidents and install new shingles in their places. To permit damaged-shingle removal, a pry bar can be used to pop out the nails fastening the damaged shingle to the roof deck and the nails holding the undamaged shingle directly above the damaged shingle which also penetrate the damaged shingle (see Fig. 6-21). The damaged shingle will then be free and can be removed. Install a new shingle where the damaged shingle was removed and nail both it and the shingle directly above in place using four roofing nails in each shingle. Covering each nail head with a little roofing cement will make the nails last longer.

Probably the first sign that shingles are close to the end of their life span is the wearing off of surface granules, which exposes black asphalt. Granules can clog gutters and cause ice dams to form in winter.

Other signs that shingles are about at the end of their life are curling edges, cracked shingles, splits, broken-off corners, which may be found on the adjacent ground after windstorms, and cracks between the shingles, especially when the cracks extend to the underlayment. Such damage may

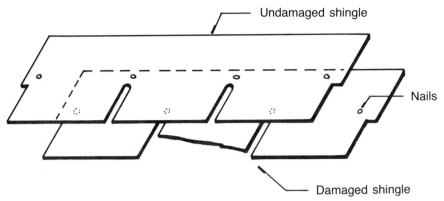

Figure 6-21 To remove the damaged shingle it is necessary to remove all the nails holding it in place.

be first apparent along the eave line. When more than 20 percent of a roof's shingles show age damage, installing new roofing is probably advisable (see Fig. 6-22).

Composition shingles that lie in constantly damp and shaded locations, especially in humid regions, may be attacked by fungi. It may be possible to remove the fungi using a pressurized water spray and a detergent wash such as is sometimes used to clean brick. Unfortunately, the process will remove some surface granules from the shingles and shorten their life. Eliminating shade from the roof may help. Where fungi is an ongoing problem, it might make sense when installing new roofing to consider using fungus-resistant shingles containing zinc or some other antifungal agent.

Nail Problems.   Damage may occur to relatively young shingles because they were improperly installed. Too few nails may have been used, for example. Or ungalvanized steel nails may have rusted through and no longer hold the shingles in place.

Where nails have failed and the shingles are otherwise in good condition, which is a highly unlikely situation, shingles may be renailed. A layer of roofing cement should be applied to each nail head.

Installing New Composition Shingles over Existing Materials

There are two options for installing new composition shingles where roofing exists: remove the existing materials or leave the existing materials in place and roof over them.

Usually, new asphalt shingles can be applied directly over properly prepared existing roll roofing or asphalt or wood shingles. Some manufacturers

Figure 6-22 Replacing old shingles with new (the lighter shingles) is sometimes a waste of money. (*Photo by author.*)

recommend that not more than two courses of composition shingles be applied on a roof. All sources agree that under no circumstances should more than three layers of composition shingles be installed on any roof.

Removing Existing Roofing.   One way to prepare an existing roof to receive new composition shingles is to completely remove the existing roofing and underlayment. That method will work regardless of the type of existing roofing, as long as the roof deck is appropriate for composition shingle application. A roof structure that supported roll roofing, composition shingles, wood shakes or shingles, slate, roofing tile, or metal roofing will probably support new composition shingles and appropriate underlayment. Of course, the roof deck must be nailable or be overlaid with a nailable material. The weight of such an overlay must be considered when determining the ability of the structure to hold the new loads.

It will usually be necessary to remove slate, tile, and metal roofing before installing composition shingles.

Roofing over Existing Materials.   Where new shingles are to be installed over existing roofing, existing underlayment usually needs no augmentation. Nonexistent metal edges and ice barriers should, however, be added.

For general recommendations for preparing existing roofing to receive new composition shingles, refer to "Preparing Existing Roofing to Receive New Roofing" in this chapter.

For general recommendations for preparing existing underlayment to receive new composition shingles, refer to "Preparing Existing Underlayment to Receive New Roofing" in this chapter.

New shingles should be laid so that they span over the butt edges of left-in-place existing shingles. New nails should be staggered to miss those in the existing roofing by at least 2 inches.

New composition shingles are often installed directly over a layer of existing composition shingles.

New composition shingles may be installed directly over existing roll roofing, providing that the slope is sufficient, the existing roofing, deck, and structure are sound and sufficiently strong, and the existing roofing is smooth.

It is possible to use asphalt shingles to reroof over left-in-place wood shingles, providing that the structure will support the loads and that the existing wood-shingled roof can be made smooth enough to receive the new composition shingles.

Every substrate to receive composition shingles must be clean, smooth, and free of protruding nails and loose impediments, dirt, and other deleterious materials.

Repairing and Extending Existing Slate Roofs

Slate roof problems are usually due to failure in flashing, guttering, or nails, or result from improper installation. Falling limbs or people walking on the roof can cause damage, of course, and slate is also susceptible to breaking in winter by ice, especially due to ice dams.

Slate roof problems are rarely due to slate failure. Old slate may delaminate or crumble due to weathering, of course. Deterioration may occur first around exposed nails used to repair slate. Alternate freezing and thawing of water entering minor erosions at nail holes contributes to breakdown. Air pollution can shorten slate life somewhat. Even so, serious deterioration is not usually a major problem in any slate less than 40 years old. In most slate, such damage should not occur for 100 years or more (see Table 6-2).

Building additions will sometimes require that new slate join existing slate.

Repairing Flashings.   Flashing problems are discussed in Chapter 8. Repairing flashings may force removing, and later reinstalling, some slate.

Repairing Roofing and Underlayments.   Underlayment beneath slate is needed to act as a cushion when the slate is installed over sheathing but has little to do with water resistance. Therefore, repairs to underlayment are seldom necessary. When they are, chances are that the slate has been removed. Repairs then amount to removing the damaged underlayment and installing a new piece, or simply nailing a new piece of felt over the existing felt.

When a slate roof shows slipped, cracked, or missing slate, look for metal clips or metal nail covers of the type mentioned in the next few paragraphs. Such devices show that previous damage has been repaired and, if excessive in number, may suggest that the original nails are no longer properly holding the slate. See "Nail and Nail Hole Problems" further along in this chapter.

Examine slating battens when slates are missing. Where exposed, examine slating battens from below. Check to see if wood members have been damaged by water intrusion. Repair or remove and install new wood battens where the existing ones are no longer usable.

Select slate for repairs from the same source as the existing slate. Where possible, obtain weathered slate so that the patch will match the existing roofing. When selecting patching slates for graduated slate roofs, the location within the roof will affect the thickness and size of slate needed.

The best method for installing a new slate to replace a damaged or missing slate is as follows (see Fig. 6-23):

1. Remove damaged slate completely after cutting supporting nails using a slate ripper (a tool designed for the purpose). Also cut the nails off where slate is missing. This is a good time to ascertain what nail material was used in the original installation. If they are corrodible nails, perhaps it is time to consider removing the slate and reinstalling it properly using copper nails.
2. Insert a new slate.
3. Nail through the joint between the covering slates and the new slate into the sheathing.
4. Bend a 3-inch-wide by about 8-inch-long copper flashing strip into a concave shape and place it concave side up beneath the covering slates to cover the new nail. Leave bottom of copper strip 2 inches below nail to protect nail from weather. Leave top of strip at least 2 inches higher than bottom of overlying shingle to prevent water from blowing up and over the strip. Use a piece longer than 8 inches if necessary.

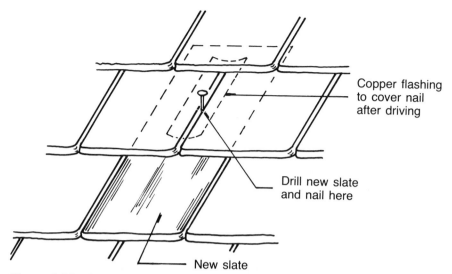

Copper flashing
to cover nail
after driving

Drill new slate
and nail here

New slate

Figure 6-23 Replacing a damaged or missing slate using a nail covered by a concave copper flashing strip.

It is also possible to place new individual slates using a copper tab method (see Fig. 6-24). The major disadvantage of this method is that the copper tabs remain visible in the completed work. The method is as follows:

1. Remove damaged slate completely by cutting supporting nails with a ripper. Also cut the nails off where slate is missing.
2. Drill or punch nail holes in the two underlying slates and nail in place two 2-inch-wide copper tabs.
3. Insert new slate and bend copper tabs over the bottom to hold the slate in place.

Nail and Nail Hole Problems.   Most failures in roofing slate result from improper nail holes, improper nailing, or use of the wrong nails.

When nails are driven too tightly, the slate may crack when the roof flexes, or the slate may rise up over the nail and be blown from the roof in high winds. Nails not driven in far enough can serve as a point support for overlying slate and cause the overlying slate to break under an applied load, such as a foot step, sometimes even by wind loads. Nails that are too short may not hold in the deck.

Another common problem is using nails that corrode. Using corrodible nails in slate roofs is a more serious problem than the same use in many other kinds of roofing materials because the slate is likely to last much

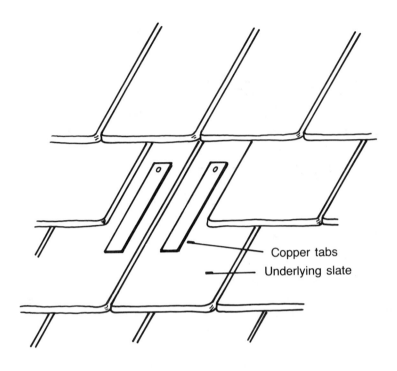

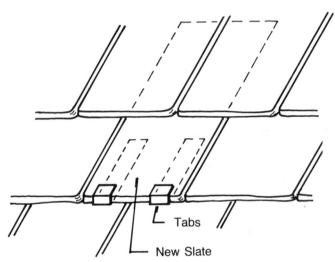

Figure 6-24 Replacing a damaged or missing slate using the copper tab method. The top drawing shows the slate removed and the copper tabs in place. The bottom shows the completed installation.

longer than the corrodible nails. Money saved by using less expensive nails will be spent many times over when rusted-through nails force relaying of the slate.

When nails fail, it is necessary to remove the affected slates and reinstall them, this time using the proper size and length of hard copper nails. When slates are letting go in large numbers (a subjective judgment), perhaps it is time to remove the entire slate roof and reinstall the slate using copper nails.

Elastic Cement Problems.   Sometimes, application of too much elastic cement in the initial installation can contribute to leaks. For example, elastic cement along open valley flashing lines can prevent water that gets beneath the shingles from escaping onto the flashing and running off. When that happens, the slate must be removed, the elastic cement removed, and the slate reinstalled.

Installing Slate Roofing at Additions

When an addition requires extending an existing slate roof, the major problem is blending the new slate with the old. It will probably not be possible to exactly match the existing slate, at least not initially. If the new slate is carefully selected, however, it may weather to match the existing slate over time.

The matching problem will often be eased somewhat if a portion of the existing slate along the juncture is removed and the new slate intermixed with the old. Such a procedure will reduce the sharpness of the juncture between new and old and make the joining less conspicuous.

Where repairs are occasioned by the addition or removal of an element, such as a dormer, the removed slate can be used to roof the new element or the opening where the element was removed. In such cases, use whatever new slate is necessary in the shadows when possible.

Installing New Slate Roofing over Existing Materials

When slate begins to delaminate or crumble, the only options are to install new slate or another roofing material. But because of the long life of slate, most slate roofing projects on existing buildings either involve reinstalling slate removed because of fastener failure or laying new slate on a roof previously roofed with less permanent materials. Slate can be laid over the existing roofing in certain circumstances or the existing material can be removed first.

Removing Existing Materials.   Completely removing the existing roofing and underlayments before new slate is installed yields the best results, especially if the existing roofing has been in place for a long time. Removing the existing roofing and underlayment permits inspection of sheathing and preparation for the new slate in the same way as would be done on a new roof. It is always better to start with an ideal substrate when installing slate. Slate itself lasts so long that it is a shame to build into a slate roof conditions that will fail before the slate does.

Another advantage of removing the existing roofing is that its weight need not be taken into account when determining if the existing roof structure will support the slate and the construction loads associated with installing the slate.

Before installing slate over an existing roof structure, even after removing the existing roofing, have a structural engineer verify the adequacy of the structure. Sometimes, a roof structure designed for less permanent roofing, such as composition shingles, will support 3/16-inch-thick commercial standard slate roofing. Roof structure designed to support wood shingles will usually support commercial standard slate. Roofs designed to carry tile will probably hold slate without a problem. Nevertheless, it is wise to verify the adequacy of every roof.

After the existing roofing has been removed and before slating begins, it should be verified that sheathing joints fall over supports. Improperly laid or loose sheathing should be renailed or removed and relaid. Unsound, broken, rotted, or otherwise unsatisfactory sheathing should be removed and new sheathing installed. Warped or raised sheathing edges should be cut or sanded down.

Before installing new underlayment, existing protruding nails should be pulled or driven down flush with the sheathing. Immediately before laying new felt, the roof deck should be swept clean and loose impediments removed.

New underlayment and slate are installed over a properly prepared existing roof deck in the same way as they would be applied on new roofs.

Roofing over Existing Materials.   Before deciding to install a new slate roof on an existing roof without removing the existing roofing, consult with a structural engineer and the slate producer about the advisability of doing so. Roofing over existing light roofing such as composition shingles, wood shingles, or metal roofing may be possible. A few roofs may be strong enough to hold existing wood shakes or mineral-fiber-cement tile and new slate. Not many roofs, however, are adequately designed to support an existing heavy roofing, such as slate or tile, and new slate as well. Even 3/16-inch-thick slate can weigh 750 pounds, or more, per square. Graduated slate roofing can weigh 7,000 or 8,000 pounds per square.

Before slate roofing is applied over existing roofing, loose materials should be securely fastened in place, low spots filled, and rough conditions eased. For general recommendations for preparing existing roofing to receive new roofing, refer to "Preparing Existing Roofing to Receive New Roofing" in this chapter. Where slate must bridge over existing wood shingle butts, each slate should span at least two courses of wood shingles. Slate placed over existing wood shingles should have four nail holes instead of two so that if good nailing is not achieved through the regular holes, which sometimes happens when nailing through existing wood shingles, additional options are available.

It sometimes makes sense to remove the existing roofing but to permit the underlayment to remain. It is necessary, however, to ascertain the condition of the substrate beneath the underlayment before deciding to leave the underlayment in place. Existing underlayment should only be left in place when it is in good condition and when the substrates are sound. If the underlayment is essentially sound but somewhat damaged, it might be a good idea to add a course of underlayment. When the existing underlayment is lighter than recommended, add a layer. For general recommendations for preparing existing underlayment to receive new roofing, refer to "Preparing Existing Underlayment to Receive New Roofing" in this chapter.

Apply new slate over existing materials using proper nails and procedures similar to those for new slate roofing. Appearance should be the same as that for new slate roofing.

Repairing Wood Shingle and Shake Roofing

Most problems with wood shingle and shake roofing are due to improper installation or failed flashing. Improper installation includes using corrodible nails or lining up joints in succeeding courses so that water can find its way between the shingles or shakes into the roof construction below.

Wood shingles can become broken from exterior-generated trauma, but most breakage is due either to aging or improper nailing. Warp, curl, and splitting are to be expected in wood shingle and shake roofing. That is part of the charm of the materials. All shingles and shakes will split, warp, and curl as they age (see Fig. 6-25). Some, unfortunately, will age faster than normal and need replacement before the entire roof reaches its expected life span. Improper nailing can make the problems appear earlier than they otherwise would. Using more than two nails per shingle or shake, for example, will increase splitting. Replacement is indicated when shingles or shakes split enough to become loose or warp and curl excessively. The time to replace them is before they blow off the roof in a windstorm or leaks show up inside.

Losing a few shingles does not, however, mean that the entire roofing has exceeded its useful life. Red cedar shingles and shakes have natural

Figure 6-25 A 25-year-old wood shake roof beginning to show its age. This roof should be good for another 5 years or so, but the owner should not be surprised if problems begin to occur sooner. (*Photo by author.*)

life spans of about 15 to 30 years, respectively, and may last even longer if properly installed and regularly maintained.

A sound wood shingle or shake roof will appear even and consistent in appearance. Wood shakes or shingle roofs in poor condition appear ragged. Shingles or shakes may be broken or missing. Units may show excessive splitting, warping, or curling. In shaded areas, shingles or shakes may show fungal growth and resultant disintegration.

Repairing Flashings.   Refer to Chapter 8 for a discussion of metal flashing. Repairing flashing may require removal of some shingles or shakes. It might be possible to reinstall carefully removed shingles or shakes, but often it will be necessary to provide new shingles or shakes.

Repairing Roofing and Underlayments.   Underlayment beneath wood shingles or shakes can be repaired if damaged, but most such damage will not be apparent until the shingles or shakes are removed. It might be possible to reuse removed underlayment and interlayer strips when installing a new roof, but it is not a good idea. Such reuse builds in a weak spot in the system, which is unwise since shingles and shakes enjoy a relatively long life span. The savings from reusing existing underlayments and interlayers is ill spent.

Where an entire roof is covered with underlayment beneath wood shingles or shakes is a different matter, however. If the underlayment is in good condition, reuse may be appropriate. For general recommendations for preparing existing underlayment to receive new roofing, refer to "Preparing Existing Underlayment to Receive New Roofing" in this chapter.

When a wood shingle or shake roof nears its life span and problems begin to appear, it is a good idea to determine if the problems are widespread or localized. Metal tabs of the type discussed in the next few paragraphs are a sign that other shingles or shakes have been replaced. Many instances of replacement coupled with a general overall ragged appearance is probably a sign that new roofing is needed. If the roof is losing shingles or shakes every time the wind blows or shingles or shakes are excessively bowed, curled, or otherwise warped, that is also a sign that it is time to replace or re-cover the roof.

A temporary repair can be made by sliding a piece of sheet metal beneath damaged or missing shingle or shakes. Temporary repairs should be removed and permanent repairs made as soon as possible.

New shingles and shakes can be installed without removing other shingles or shakes in two ways. The best way to replace the damaged shingles or shakes is as follows (see Fig. 6-26).

1. Split the damaged shingle or shake with a chisel and remove the pieces.
2. Cut the nails with a slate ripper or hacksaw.

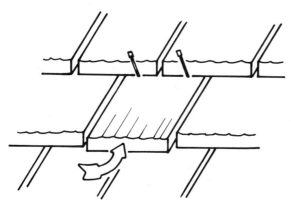

Figure 6-26 Replacing a damaged or missing wood shingle or shake. The nails should be hidden in the completed repair.

3. Slide a new shake or shingle into place with the butt protruding 1/4 inch below the adjacent shingles or shakes.
4. Drive two nails through the new shingle into the sheathing at the line of the covering shingles. Drive these nails at an angle (toenail) into the shingle near the top.
5. Drive the new shingle or shake home using a board to protect the shingles or shakes from damage, thus driving the nails up beneath the overlying shingles or shakes.

Another effective way to install a new wood shingle or shake is as follows (see Fig. 6-27):

1. Split the damaged shingle or shake with a chisel and remove the pieces.
2. Cut the nails with a slate ripper or hacksaw.
3. Using copper roofer's nails, nail in place a one-inch-wide copper tab.
4. Insert a new shingle or shake and bend the copper tab over the bottom to hold the shingle or shake in place.

The disadvantage of the second method is that the metal tabs are visible in the complete work.

Installing New Wood Shingles and Shakes over Existing Materials

There are two options for installing new wood shingles or shakes where there is existing roofing: remove the existing materials or leave the existing materials in place and roof over them.

Considerations for deciding whether to leave the existing roofing in place or remove it are discussed in "Repair, Replacement, and Re-Covering Work—General Requirements."

Removing Existing Materials.   One way to prepare an existing roof to receive new wood shingles or shakes is to completely remove the existing roofing and underlayment. This method will work regardless of the type of existing roofing, as long as the decking is appropriate (nailable) for wood shingle or shake installation. A roof structure that supported wood shingles or shakes, slate, or tile will almost certainly support new wood shingles or shakes. A deck that supported composition shingles might support wood shingles or shakes or might support shingles but not shakes.

Roofing over Existing Materials.   Wood shingles and shakes cannot be applied over existing slate or tile, but where the roof structure is sufficiently

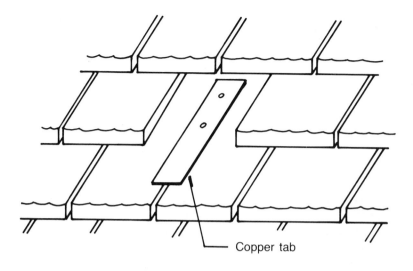

Copper tab

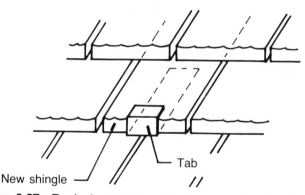

Tab

New shingle

Figure 6-27 Replacing a wood shingle or shake using the copper tab method. The top drawing shows the wood removed and the copper tab in place. The bottom shows the completed installation.

strong and the existing material is in good enough condition, they can be applied directly over most other steep roofing materials.

For general recommendations for preparing existing roofing to receive new roofing, refer to "Preparing Existing Roofing to Receive New Roofing" in this chapter.

Where existing wood shingles or shakes are too rough for conventional preparation, it might be possible to apply strip sheathing over them to accept the new wood shingles or shakes. Of course, such a solution adds more weight, which should be considered when deciding whether the existing roof's structural capability is sufficient.

Metal roofing or flashings to be left in place beneath wood shingles or shakes should be prepared in accordance with the recommendations of the Red Cedar Shingle and Handsplit Shake Bureau as applicable to the particular metal material. It is not a good idea to install wood shingles or shakes over a copper roof or to leave copper flashing in place where it will contact wood shingles or shakes. Rainwater can leach tannic acid from red cedar, which corrodes copper. Other metals may also be susceptible. If the flashing or roofing is other than stainless steel, aluminum, or galvanized metal, contact the Red Cedar Shingle and Handsplit Shake Bureau for advice.

After proper preparation, new wood shingles and shakes should be laid and fastened in place using nails long enough to penetrate the sheathing. Appearance should be that of a new roof.

Repairing Clay Tile Roofing

Clay tile roof problems are usually due to failure in flashing, guttering, or nails, or result from improper installation. Falling limbs or people walking on the roof can cause damage, of course, and clay tile is also susceptible to breaking in winter by ice, especially due to ice dams.

In any clay tile roof, some tiles will be inferior and will disintegrate or crumble before the life span of the roof has been reached. But serious deterioration is not usually a major problem in clay tile roofing less than 50 years old. Often, such damage will not occur for 100 years or more (see Table 6-2).

Building additions will sometimes require that new clay tile join existing clay tile.

Repairing Flashings.   Flashing problems are discussed in Chapter 8. Repairing flashings may necessitate removing, and later reinstalling, some clay tile.

Repairing Roofing and Underlayments.   Underlayment beneath clay tile can be repaired if damaged, but most such damage will not be apparent until the clay tile has been removed for some other reason. Repairs, then, amount to removing the damaged underlayment and installing a new piece, or simply nailing a new felt layer over the existing felt.

When a roof shows slipped, cracked, or missing clay tile, look for metal tabs of the type mentioned in the next few paragraphs. Such devices show that previous damage has been repaired and suggests that the original fasteners are no longer holding the clay tile. See "Fastener Problems" later in this chapter.

Examine battens, if any, when clay tiles are missing. Where exposed, examine battens from below. Check to see if wood members have been damaged by water intrusion. Repair or remove and install new wood battens where the existing ones are no longer usable.

Select clay tile for repairs from the same source as the existing clay tile. Where possible, obtain weathered clay tile so that the patch will match the existing roofing.

Replacing a damaged or missing clay tile is sometimes relatively easy where battens have been used in the existing roof. First, remove the remains of the existing tile and cut the nails that held it in place and then lift the tile immediately above the space where the new tile will be installed just enough so that the new tile can be slid beneath it to the level of the battens. Hook the new tile over the batten and lower the raised overlying tile back into proper position; the installation is complete. The method will not work, of course, if the tile used has no lugs for hooking over battens.

It is also sometimes possible to place new individual clay tiles using a copper tab method. The major disadvantage of this method is that the copper tabs remain visible in the completed work. The method is as follows:

1. Remove damaged clay tile completely by cutting supporting nails with a ripper. Also cut the nails off where clay tile is missing.
2. Nail in place, using noncorrosive roofers nails, a 2-inch-wide heavy copper tab, following a method similar to that shown in Figure 6-27. Use a double-thickness tab (see Fig. 6-28). In addition, it might be a good idea to test the tab to ensure that it will be strong enough to support the weight of the tile and loads to be applied, such as snow load. If they have experience with such installations, tile manufacturer's or roof repair contractor's recommendations may be reliable.

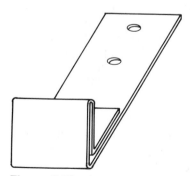

Figure 6-28 A copper tab for heavy materials.

For some tile shapes, it may be necessary to install a raised nailer beneath the tile to support the tabs at the correct elevation.

3. Insert new clay tile and bend copper tabs over the bottom to hold the clay tile in place.

Fastener Problems.   Many clay tile roof failures result from improper fastener installation or use of the wrong fasteners.

When nails are driven too tightly, clay tile may crack when the roof flexes. Overdriven nails will also raise the butt of the clay tile, allowing water to be blown over the headlap and letting high winds get beneath the tile and blow it off the roof. Nails not driven in far enough can serve as a point support for overlying clay tile and cause the overlying clay tile to break under an applied load, such as a footstep, sometimes even by wind loads. Nails that are too short may not hold in the deck. Smooth shank nails may pull out of the substrates. Any nail may pull out of plywood after a time, which leads some roofing experts to warn against nailing roofing tile down to plywood under any circumstances.

Another common problem is using fasteners that corrode (see Fig. 6-29). Using corrodible fasteners in clay tile roofs is a more serious problem than the same use in many other kinds of roofing materials because the clay tile is likely to last much longer than the corrodible fasteners. Money saved by using less expensive fasteners will be spent many times over when rusted through fasteners force re-laying the clay tile.

When fasteners fail, it is necessary to remove the affected clay tiles and reinstall them, this time using the proper size and length of hard copper nails. When clay tiles are letting go in large numbers (a subjective judgment), perhaps it is time to remove the entire clay tile roof and reinstall the tile using copper nails.

Elastic Cement and Mortar Problems.   Sometimes, application of too much elastic cement or mortar in the initial installation can contribute to leaks. For example, elastic cement or mortar along open valley flashing lines can prevent water that gets beneath the tile from escaping onto the flashing and running off. When such a thing happens, the clay tile must be removed, the elastic cement must be removed, and the clay tile can be reinstalled.

Installing Clay Tile Roofing at Additions

When an addition requires extending an existing clay tile roof, a major problem is blending the new clay tile with the old. It will probably not be possible to exactly match the existing clay tile, at least not initially. If the

Figure 6-29 Nongalvanized steel nail failure was the real cause of this wind damage. Tile on the entire roof must be relayed. (An odd aside here: The nails in the underlayment, which can be seen clearly in the lower photo, were galvanized.) (*Photos by author.*)

new clay tile is carefully selected, however, it may weather to match the existing over time.

The matching problem will often be eased somewhat if a portion of the existing clay tile along the juncture is removed and the new clay tile intermixed with the old. Such a procedure will reduce the sharpness of the juncture between new and old and make the joining less conspicuous.

Where repairs are occasioned by addition or removal of an element, such as a dormer, the removed tile can be used to roof the new element or the opening where the element was removed. Then, use whatever new clay tile is necessary in the shadows when possible.

Installing New Clay Tile Roofing over Existing Materials

When clay tile begins to disintegrate or crumble, the only options are to install new clay tile or another roofing material. But because of the long life of clay tile, most clay tile roofing projects on existing buildings involve either re-laying tile whose fasteners have failed or laying new clay tile on a roof previously roofed with less permanent materials. Clay tile can be laid over the existing roofing in certain circumstances or the existing material can be removed first. Refer to "Repair, Replacement, and Re-Covering Work—General Requirements" for a discussion of the factors involved in deciding whether to remove the existing roofing or roof over it.

Removing Existing Materials.   Completely removing the existing roofing and underlayments before new clay tile is installed yields the best results, especially if the existing roofing has been in place for a long time. Removing the existing roofing and underlayment permits inspection of sheathing and preparation for the new clay tile in the same way as would be done on a new roof. It is always better to start with an ideal substrate when installing clay tile. Clay tile itself lasts so long that it is a shame to build into a clay tile roof conditions that will fail before the clay tile does.

Another advantage of removing the existing roofing is that its weight need not be taken into account when determining if the existing roof structure will support the clay tile and the construction loads associated with installing the clay tile.

Before installing clay tile over existing roof structure, even after removing the existing roofing, have a structural engineer verify the structure's adequacy. Some roof structures initially roofed with lighter materials, such as composition shingle, wood shingle or shake, metal, or mineral-fiber-cement roofing will have sufficient safety factors designed in to support clay tile roofing. Roofs designed to carry slate will usually carry clay tile, but seldom will clay tile be substituted for an existing slate roof.

After the existing roofing has been removed, and before clay tile installation begins, it should be verified that sheathing joints fall over supports. Improperly laid or loose sheathing should be renailed or removed and re-laid. Unsound, broken, rotted, or otherwise unsatisfactory sheathing should be removed and new sheathing installed. Warped or raised sheathing edges should be cut or sanded down.

Before installing new underlayment, existing protruding nails should be pulled or driven down flush with the sheathing. Immediately before laying new felt, the roof deck should be swept clean and loose impediments removed.

New underlayment and clay tile are installed over a properly prepared existing roof deck in the same way as they would be applied on new roofs.

Roofing over Existing Materials.   Before deciding to install a new clay tile roof on an existing roof without removing the existing roofing, consult with a structural engineer and the clay tile producer about the advisability of doing so. Roofing over existing light roofing such as composition shingles, wood shingles, or metal roofing may be possible. A few roofs may be strong enough to hold existing wood shakes or mineral-fiber-cement tile and new clay tile. Not many roofs, however, are adequately designed to support an existing heavy roofing, such as slate, and new clay tile as well. Even flat clay tile can weigh 800 pounds, or more, per square. Some clay tile will weigh as much as 1,400 pounds per square.

It sometimes makes sense to remove the existing roofing but to permit the underlayment to remain. It is necessary, however, to ascertain the condition of the substrate beneath the underlayment before deciding to leave the underlayment in place. Existing underlayment should only be left in place when it is in good condition and when the substrates are sound. When the existing underlayment is lighter than recommended, add a layer. For general recommendations for preparing existing underlayment to receive new roofing, refer to "Preparing Existing Underlayment to Receive New Roofing" in this chapter.

For general recommendations for preparing existing roofing to receive new roofing, refer to "Preparing Existing Roofing to Receive New Roofing" in this chapter.

Before roofing is applied over existing roofing, or existing underlayment, loose materials should be securely fastened in place, low spots filled, and rough conditions eased. Underlayment should be patched where damaged or covered with an additional felt layer. Then the existing roofing or underlayment should be stripped with new wood battens to receive the clay tile.

Apply new clay tile over new wood battens using proper nails and procedures which are the same as those recommended for clay tile installation over battens on a new roof. Appearance should be the same as that of new clay tile roofing.

Repairing Concrete Tile Roofing

Concrete tile roof problems are usually due to failure in flashing, guttering, or nails, or result from improper installation. Falling limbs or people walking on the roof can also cause damage, of course. Concrete tile is also susceptible to breaking in winter by ice, especially due to ice dams, and from other causes (see Fig. 6-30).

In any concrete tile roof, some tiles will be inferior and will spall, exfoliate, disintegrate, or crumble before the life span of the roof has been reached. But serious deterioration is not usually a major problem in concrete tile roofing less than 50 years old. Often, such damage will not even begin to occur for 75 years or more (see Table 6-2).

Building additions will sometimes require that new concrete tile join existing concrete tile.

Figure 6-30 This concrete mansard tile was struck by an object projecting from a truck. Replacement will be relatively easy because the tile is hooked over battens. (*Photo by author.*)

Repairing Flashings.   Flashing problems are discussed in Chapter 8. Repairing flashings may necessitate removing, and later reinstalling, some concrete tile.

Repairing Roofing and Underlayments.   Underlayment beneath concrete tile can be repaired if damaged, but most such damage will not be apparent until the concrete tile has been removed for some other reason. Repairs, then, amount to removing the damaged underlayment and installing a new piece, or simply nailing a new felt layer over the existing felt.

When a concrete tile roof shows slipped, cracked, or missing concrete tile, look for metal tabs of the type mentioned in the next few paragraphs or obviously new tiles. Such devices and new tiles show that previous damage has been repaired and, if excessive in number, may suggest that the original fasteners are no longer properly holding the concrete tile. See "Fastener Problems" later in this chapter.

Examine battens when concrete tiles are missing. Where possible, examine battens from below. Check to see if wood members have been damaged by water intrusion. Repair or remove and install new wood battens where the existing are no longer usable.

Select concrete tile for repairs from the same source as the existing concrete tile. Where possible, obtain weathered concrete tile so that the patch will match the existing roofing.

Replacing a damaged or missing concrete tile is often relatively easy where battens have been used in the existing roof. First, remove the remains of the existing tile and cut the nails that held it in place and then lift the tile immediately above the space where the new tile will be installed just enough so that the new tile can be slid beneath it to the level of the battens. Hook the new tile over the batten and lower the raised overlying tile back into proper position; the installation is complete. The method will not work, of course, if the tile used has no lugs for hooking over battens. Fortunately, most concrete tile has batten lugs.

Some experts believe that there is no good way to install a replacement concrete tile on roofs where battens were not used. It is sometimes possible to place new individual concrete tiles using a copper tab method. The major disadvantage of this method is that the copper tabs remain visible in the completed work. In addition, this method may not work with some of the heavier types of concrete tile, which can weigh more than 10 pounds each. The method is as follows:

1. Remove damaged concrete tile completely after breaking it up with a hammer. Then cut and remove supporting nails. Also cut the nails off where concrete tile is missing.
2. Nail in place, using noncorrosive roofers nails, a 2-inch-wide heavy

copper tab, following a method similar to that shown in Figure 6-27. Use a double thickness tab (see Fig. 6-28). It might be well to test the tab to ensure that it will be strong enough to support the weight of the tile and loads to be applied, such as snow load. If they have experience with such installation, the tile manufacturer's or roof repair contractor's recommendations may be reliable. For barrel tile, it may be necessary to install a raised nailer beneath the tile to support the tabs at the correct elevation.

3. Slide a new concrete tile into place and bend the copper tab over the bottom to hold the concrete tile in place.

It might be possible to wire tie a new concrete tile into an opening left by a removed tile. The method is to drill the adjacent existing tiles and the new tiles through the overlapping lugs or other concealed portions and wire the two together, pulling the wire tight as the new tile is slid into place. The wire ends are then shoved into the joint between the tiles to conceal them. This is a tricky maneuver, requiring expert hands.

Fastener Problems.   Many concrete tile roof failures result from improper fastener installation or using the wrong fasteners.

When nails are driven too tightly, concrete tile may crack when the roof flexes. Overdriven nails will also raise the butt of the concrete tile, allowing water to be blown over the headlap and letting high winds get beneath the tile and blow it off the roof. Nails not driven in far enough can serve as a point support for overlying concrete tile and cause the overlying concrete tile to break under an applied load, such as a footstep, and sometimes even by wind loads. Nails that are too short may not hold in the deck. Smooth shank nails may pull out of the substrates. Any nail may pull out of plywood after a time, which leads some roofing experts to urge against nailing roofing tile down to plywood under any circumstances.

Another common problem is using fasteners that corrode. Using corrodible fasteners in concrete tile roofs is a more serious problem than the same use in many other kinds of roofing materials because the concrete tile is likely to last much longer than the corrodible fasteners. Money saved by using less expensive fasteners will be spent many times over when rusted through fasteners force re-laying the concrete tile.

When fasteners fail, it is necessary to remove the affected concrete tiles and reinstall them, this time using the proper size and length of noncorrosive nails. When concrete tiles are letting go in large numbers (a subjective judgment), perhaps it is time to remove the entire concrete tile roof and reinstall the concrete tile using noncorrosive nails.

Elastic Cement and Mortar Problems.   Sometimes, application of too much elastic cement or mortar in the initial installation can contribute to leaks. For example, elastic cement or mortar along open valley flashing lines can prevent water that gets beneath the tile from escaping onto the flashing and running off. When such a thing happens, the concrete tile must be removed, the elastic cement removed, and the concrete tile can be reinstalled.

Installing Concrete Tile Roofing at Additions

When an addition requires extending an existing concrete tile roof that is not field-painted, blending the new concrete tile with the old can be a major problem. It will probably not be possible to exactly match the existing concrete tile, at least not initially. If the new concrete tile is carefully selected, however, it may weather to match the existing over time.

The matching problem will often be eased somewhat if a portion of the existing concrete tile along the juncture is removed and the new concrete tile intermixed with the old. Such a procedure will reduce the sharpness of the juncture between new and old and make the joining less conspicuous.

Where repairs are occasioned by the addition or removal of an element, such as a dormer, the removed concrete tile can be used to roof the new element or the opening where the element was removed. In such cases, use whatever new concrete tile is necessary in the shadows when possible.

Installing New Concrete Tile Roofing over Existing Materials

When concrete tile begins to spall, exfoliate, disintegrate, or crumble due to age, the only options are to install new concrete tile or another roofing material. But because of the long life of concrete tile, most concrete tile roofing projects on existing buildings are either to re-lay tile whose fasteners have failed or to lay new concrete tile on a roof previously roofed with less permanent materials. Concrete tile can be laid over the existing roofing in certain circumstances or the existing material can be removed first. Refer to "Repair, Replacement, and Re-Covering Work—General Requirements" for a discussion of the factors involved in deciding whether to remove the existing roofing or roof over it.

Removing Existing Materials. Completely removing the existing roofing and underlayments before new concrete tile is installed yields the best results, especially if the existing roofing has been in place for a long time. Removing the existing roofing and underlayment permits inspection of sheathing and preparation for the new concrete tile in the same way as would be done on a new roof. It is always better to start with an ideal substrate when installing concrete tile. Concrete tile itself lasts so long that it is a shame to build into a concrete tile roof conditions that will fail before the concrete tile does.

Another advantage of removing the existing roofing is that its weight need not be taken into account when determining if the existing roof structure will support the concrete tile and the construction loads associated with installing the concrete tile.

Before installing concrete tile over existing roof structure, even after removing the existing roofing, have a structural engineer verify the adequacy of the structure. Some roof structures initially roofed with lighter materials, such as composition shingle, wood shingle or shake, metal, or mineral-fiber-cement roofing, will have sufficient safety factors designed in to support concrete tile roofing. Roofs designed to carry slate will usually carry concrete tile, but seldom will concrete tile be substituted for an existing slate roof.

After the existing roofing has been removed, and before concrete tile installation begins, it should be verified that sheathing joints fall over supports. Improperly laid or loose sheathing should be renailed, or removed and re-laid. Unsound, broken, rotted, or otherwise unsatisfactory sheathing should be removed and new sheathing installed. Warped or raised sheathing edges should be cut or sanded down.

Before installing new underlayment, existing protruding nails should be pulled or driven down flush with the sheathing. Immediately before laying new felt, the roof deck should be swept clean and loose impediments removed.

New underlayment and concrete tile are installed over a properly prepared existing roof deck in the same way as they would be applied on new roofs.

Roofing over Existing Materials.   Before deciding to install a new concrete tile roof on an existing roof without removing the existing roofing, consult with a structural engineer and the concrete tile producer about the advisability of doing so. Roofing over existing light roofing such as composition shingles, wood shingles, or metal roofing may be possible. A few roofs may be strong enough to hold existing wood shakes or mineral-fiber-cement tile and new concrete tile. Not many roofs, however, are adequately designed to support an existing heavy roofing, such as slate, and new concrete tile as well. Even lightweight concrete tile will weigh 800 pounds, or more, per square. Some concrete tile will weigh as much as 1,300 pounds per square.

It sometimes makes sense to remove the existing roofing but to permit the underlayment to remain. It is necessary, however, to ascertain the condition of the substrate beneath the underlayment before deciding to leave the underlayment in place. Existing underlayment should only be left in place when it is in good condition and when the substrates are sound. When the existing underlayment is lighter than recommended, add a layer.

For general recommendations for preparing existing underlayment to receive new roofing, refer to "Preparing Existing Underlayment to Receive New Roofing" in this chapter.

For general recommendations for preparing existing roofing to receive new roofing, refer to "Preparing Existing Roofing to Receive New Roofing" in this chapter.

Before roofing is applied over existing roofing, or existing underlayment, loose materials should be securely fastened in place, low spots filled, and rough conditions eased. Underlayment should be patched where damaged or covered with an additional felt layer. Then the existing roofing or underlayment should be stripped with new wood battens to receive the concrete tile.

Apply new concrete tile over new wood battens using proper nails and procedures the same as those recommended for concrete tile installation over battens on a new roof. Appearance should be the same as that of new concrete tile roofing.

Repairing Mineral-Fiber-Cement Tile

Mineral-fiber-cement roofing problems are often due to failure in flashing, guttering, or fasteners, or result from improper installation. Mineral-fiber-cement tiles are brittle and can be broken by falling tree limbs, people walking on the roof, or ice, especially where ice dams form beneath the tile. Improper installation can include using corrodible fasteners, improperly driving fasteners, or lining up joints in succeeding courses so that water can find its way between the tiles into the roof construction below.

Mineral-fiber-cement roof problems are sometimes due to failure of the material itself brought on by aging. When old mineral-fiber-cement tile begins to disintegrate, providing new tile is the only solution. Fortunately, deterioration solely because of age will seldom occur in less than 25 years and usually does not happen for 40 years. Some mineral-fiber-cement tile roofs have been in use for more than 50 years with no major problems. In fact, some manufacturers today warrant their product for 50 years.

Broken tiles should be removed and new tiles placed before they blow off the roof in a windstorm or leaks show up inside. But losing a few shingles on a roof that has not reached its expected life span is not a serious problem, unless the loss is due to corroded fasteners.

As mineral-fiber-cement tiles age, they absorb more and more moisture. In time, the tiles may show moss and other organic growth, which may be unsightly but is not harmful unless it grows thick enough to block water flow or contribute to ice buildup. Excessive organic growth can be removed by hand scraping.

Repairing Flashings.   Metal flashing problems are discussed in Chapter 8. Repairing flashing may require removal of some tile. It might be possible to reinstall carefully removed tile, but often it will be necessary to provide new tile.

Repairing Roofing and Underlayments.   When repairs to continuous underlayment are necessary, the mineral-fiber-cement tile has usually been removed. Repairs, then, amount to removing the damaged underlayment and installing a new piece or fastening a new piece of felt over the existing felt.

It might be possible to reuse removed underlayment and interlayer strips when installing a new roof, but it is not a good idea. Such reuse builds a weak link into the system, which is unwise since mineral-fiber-cement tile roofing enjoys a relatively long life span. The savings from reusing existing underlayments and interlayers is ill spent.

Where an entire roof is covered with underlayment beneath mineral-fiber-cement tile and the underlayment is in good condition, reuse may be appropriate. Small damages can be covered with a new felt layer.

When a mineral-fiber-cement tile roof nears its life span and problems begin to appear, it is a good idea to determine if the problems are widespread or localized. Metal tabs of the type discussed in the next few paragraphs are a sign that other tiles have been replaced. Many instances of replacement coupled with a tendency for tiles to blow off in every windstorm is probably a sign that new roofing is needed.

When a mineral-fiber-cement tile roof that is not near its expected life span shows slipped, cracked, or missing tiles, and many metal tabs or metal nail covers exist, this situation may suggest that the original fasteners are no longer properly holding the tile. See "Fastener Problems" later in this chapter.

Examine sheathing when tiles are missing. Where possible, examine sheathing from below. Check to see if wood members have been damaged by water intrusion. Repair or remove and install new sheathing where the existing is no longer usable.

A temporary repair can be made by sliding a piece of sheet metal or asphalt roll roofing beneath damaged tile or in an opening left by missing tiles and painting the patch to match the adjacent roofing. Temporary repairs should be removed and permanent repairs made as soon as possible.

Select tile for repairs from the same source as the existing tile, where possible, in order to match the existing roofing.

The best method for installing a new mineral-fiber-cement tile to replace a damaged or missing tile is to use one similar to that shown in Figure 6-23 and described as follows:

1. Remove damaged tile completely after cutting supporting fasteners using a slate ripper or hacksaw. Also cut the fasteners off where tile is missing. This is a good time to determine if the fasteners used in the original installation were corrodible.
2. If the decision has been made to not replace or re-cover the existing roofing at the time, insert a new tile.
3. Nail through the joint between the covering tile and the new tile into the sheathing.
4. Bend a 3-inch-wide by about 8-inch-long copper flashing strip into a concave shape and place it concave side up beneath the covering tile to cover the new nail. Leave bottom of copper strip 2 inches below nail to protect nail from weather. Leave top of strip at least 2 inches higher than bottom of overlying tile to prevent water from blowing up and over the strip. Use a piece longer than 8 inches if necessary.

It is also possible to place new individual tiles using a copper tab method similar to that shown in Figure 6-24. The major disadvantage of this method is that the copper tabs remain visible in the completed work. The method is as follows:

1. Remove damaged tile completely by cutting supporting fasteners with a ripper or hacksaw. Also cut the fasteners off where tile is missing.
2. Drill nail holes in the two underlying tiles and nail in place two 2-inch-wide copper tabs.
3. Insert new tiles and bend copper tabs over the bottom to hold the tiles in place.

Fastener Problems.   Most mineral-fiber-cement roofing failures result from improper fastening or the use of the wrong fasteners.

When fasteners are driven too tightly, the tile may crack when the roof flexes, or the tile may rise up over the fastener and be blown from the roof in high winds. Fasteners not driven in far enough can serve as a point support for overlying tile and cause the overlying tile to break under an applied load, such as a footstep, sometimes even by wind loads. Fasteners that are too short may not hold in the deck.

Another common problem is using fasteners that corrode, which is false economy because the tile will last longer than the fasteners. Money saved by using less expensive fasteners will be spent several times over when rusted through fasteners force re-laying the tile or providing new tile prematurely.

When fasteners fail, it is necessary to remove the affected tiles and reinstall them, this time using the proper size, length, and type fasteners.

When tiles are letting go in large numbers (a subjective judgment), perhaps it is time to remove the entire mineral-fiber-cement roofing and reinstall the tile using galvanized fasteners.

Installing New Mineral-Fiber-Cement Tile Roofing over Existing Materials

There are two options for installing new mineral-fiber-cement tile where there is existing roofing: remove the existing materials or leave the existing materials in place and roof over them.

Refer to "Repair, Replacement, and Re-Covering Work—General Requirements" for a discussion of the factors involved in deciding whether to remove the existing roofing or roof over it.

Removing Existing Materials.   One way to prepare an existing roof to receive new mineral-fiber-cement tile is to completely remove the existing roofing and underlayment. This method will work regardless of the type of existing roofing, as long as the decking is appropriate (nailable) for mineral-fiber-cement tile installation. A roof structure that supported wood shingles or shakes, slate, or tile will almost certainly support new mineral-fiber-cement tile. A roof designed for composition shingles or metal tile may support mineral-fiber-cement tile.

Roofing over Existing Materials.   Mineral-fiber-cement tile probably cannot be applied over existing slate, clay tile, or concrete tile. Not only will the combination probably be too heavy for the roof structure to support, but it is not practicable to nail through those hard materials. In addition, the heavy materials are often irregular, and making them flat enough for mineral-fiber-cement tile is too expensive to justify and will exacerbate the weight problem.

Where the roof structure is sufficiently strong and the existing material is in good enough condition, mineral-fiber-cement tile can be applied directly over roll roofing, composition shingles, wood shingles or shakes, or existing mineral-fiber-cement tile.

For general recommendations for preparing existing underlayment to receive new roofing, refer to "Preparing Existing Underlayment to Receive New Roofing" earlier in this chapter.

For general recommendations for preparing existing roofing to receive new roofing, refer to "Preparing Existing Roofing to Receive New Roofing" in this chapter.

Where existing wood shingles or shakes are too rough for normal preparation, it might be possible to apply strip sheathing over them to accept the new mineral-fiber-cement tile. Of course, such a solution adds more weight, which should be taken into account when determining the existing roof's structural capability.

After proper preparation, new mineral-fiber-cement tile should be laid and fastened in place with fasteners long enough to penetrate the sheathing. Appearance should be that of a new roof.

Repairing Metal Tile Roofing

There is no reason for metal tile to fail within its guaranteed life span if properly maintained. Most metal tile sold today is factory-prefinished, so it does not need routine painting. An older roof that was not factory-prefinished needs only periodic painting to maintain it in good condition.

At some point the finish on a metal roof will begin to fade in color. Eventually, the finish will wear thin enough for the metal to begin to show through. But most finishes are guaranteed for 20 years, and often last much longer. Some finishes are guaranteed for 35 years. Even when paint or coatings wear off, metal tile can be painted and continue in use.

Metal tile can be damaged by external trauma, of course, but most metal tile failures are due to faulty design of the tile or improper installation. Sometimes, the only solution to those problems is to remove the faulty material or incorrectly installed tile and start all over again, using new materials.

Repairing Flashings. Flashings associated with metal shingles are usually metal and often the same metal finished in the same way as the tile. Repair is discussed in Chapter 8.

Repairing Roofing and Underlayments. Underlayments will need attention only when existing metal tile roofing is removed. The method and extent of necessary repair will depend on the condition of the underlayment and the new roofing type to be installed. Underlayment in good condition may need no repairs. Minor damage can be repaired with patches. At worst, a new layer of felt may be necessary.

It is often difficult, and sometimes impossible, to remove a small portion of an existing metal tile roof and reinstall it, or even to install new tile in its place, because of the nature of the interlocking system used in the metal tile.

It is sometimes possible to patch damaged metal tile using a soldered patch. Often, all that can be done is to install a patch using fasteners and roofing cement. Refer to Chapter 8 for a discussion of techniques for patching sheet metal.

It is, of course, necessary to paint a patch in an attempt to make it match the existing tile. Matching is seldom completely successful.

Fastener Problems.   Metal tile can experience difficulties if improper fasteners are used. Improper fasteners may be incompatible with the metal of the tile, too small, or too far apart, or they may be the wrong type (straight shank nails where screws should have been used, for example).

As is true with most other types of roofing, problems associated with using incorrect fasteners or incorrectly installing fasteners is usually correctable only by removing the roofing and providing new roofing.

Installing New Metal Tile Roofing over Existing Materials

One way to prepare an existing roof to receive new metal tile is to completely remove the existing roofing and underlayment. This method will work regardless of the type of existing roofing, as long as the decking is appropriate (nailable) for metal tile installation. A roof structure that supported almost any other type of sloped roofing material will probably support new metal tile roofing. Depending on the material and tile design, metal tile roofing may weigh between 35 and 150 pounds per square of roof, which is considerably less than most other kinds of steep roofing.

Roofing over Existing Materials.   Where the roof structure is sufficiently strong and the existing material is in good enough condition, metal tile can be applied directly over any type of existing roofing that can be made reasonably flat and which will permit fasteners to be driven to the sheathing.

Refer to ''Preparing Existing Roofing to Receive New Roofing'' for a discussion of preparation methods for various existing roofing materials and underlayments.

After proper preparation, new metal tile should be laid and fastened in place using fasteners long enough to penetrate the sheathing. Appearance should be that of a new roof.

Where to Get More Information

The National Roofing Contractors Association's 1986 *The NRCA Roofing and Waterproofing Manual* includes *The NRCA Steep Roofing Manual* which has recommendations and details for applying:

- Underlayment for roll roofing, composition shingles, slate, wood shingles and shakes, and clay and concrete tile
- Metal drip edges for roll roofing and composition shingles
- Roll roofing
- Composition shingles on new surfaces
- Composition shingles over existing roll roofing, asphalt shingles, wood shingles, and bituminous underlayment

- Slate roofing on new surfaces
- Wood shingles and shakes over new surfaces
- Wood shingles and shakes over existing materials
- Wood shakes on low slope roofs
- Clay tile on steep roofs
- Concrete tile on steep roofs

Much of the slate roofing data in *The NRCA Steep Roofing Manual* derives from the National Slate Association's 1926 book *Slate Roofs*.

Much of the wood shingle and shake data in *The NRCA Steep Roofing Manual* is based on Red Cedar Shingle and Handsplit Shake Bureau recommendations.

The National Tile Roofing Manufacturers Association, Inc. (NTRMA) does not agree with *The NRCA Steep Roofing Manual* about tile roofing installation and has published a pamphlet *Installation Manual for Concrete Tile Roofing* to indicate the methods it believes are correct. NTRMA's recommendations are close to those of the manufacturers. As is pointed out in the text of this book, some of NRCA's recommendations are quite unlike those of many concrete tile manufacturers. The NTRMA pamphlet is available from NTRMA at the address listed in the Appendix under National Tile Roofing Manufacturers Association, Inc.

It would also be advisable to check with NTRMA for its latest recommendations about clay tile, which also may not agree with NRCA's recommendations.

The NRCA Steep Roofing Manual does not discuss installation of roll roofing, slate, clay tile, or concrete tile over existing construction or repairs to existing roll roofing, slate, clay tile, or concrete tile.

Ramsey/Sleeper, the AIA Committee on Architectural Graphic Standards' 1981 *Architectural Graphic Standards* contains data about roll roofing, underlayment for composition shingles, composition shingles, sheathing and underlayment for wood shakes and shingles, wood shakes and shingles, and clay tile roofing. It does not cover roofing over existing materials with roll roofing, composition shingles, slate, or tile, or repairs to existing roll roofing, composition shingles, slate, wood shakes or shingles, or tile roofing materials. It does have some data about roofing over existing roofing with wood shakes. It is a good idea to review the latest edition of *Architectural Graphic Standards* before doing new work related to the types of steep roofs covered there, but the 1981 edition contains little that will help solve existing roofing problems.

Much of the slate roofing data in the 1981 *Architectural Graphic Standards* is based on data in the National Slate Association's 1926 book *Slate Roofs*.

Much of the wood shingle and shake data in the 1981 *Architectural*

Graphic Standards is based on Red Cedar Shingle and Handsplit Shake Bureau recommendations.

The National Slate Association book *Slate Roofs* is an excellent reference for slate roofing work. It appears to have been an early source of much of the slate data in several other current authoritative references, including *The NRCA Steep Roofing Manual* and the 1981 *Architectural Graphic Standards*. *Slate Roofs* was reprinted in 1977 by Vermont Structural Slate Co., Inc., Fair Haven, VT 05743, and may still be available from them. Anyone doing remedial or new work with slate roofing should find a copy and keep it handy.

AIA Service Corporation's *Masterspec* 11/85 Basic: Section 07311, "Asphalt Shingles," and the 11/85 Basic: Section 07317, "Wood Shingles and Shakes," and most federal government guide specifications covering those same subjects include underlayment installation instructions and specifications for shingle and shake installation. The 11/85 *Masterspec* Basic: Sections 07311 and 07317, as is true for most guide specifications available today, do not, however, contain the details necessary to specify installation of shingles or shakes over existing materials, or to repair existing materials.

Masterspec's Basic Version as of 1988 does not include guide sections for specifying slate or tile roofing.

The General Services Administration's February 1970 version of Public Building Services Guide Specifications, Section 0751, "Bituminous Roof Repair" (which is applicable to roll roofing) and Section 0755, "Repair of Built-Up Base Flashing" cover repair of existing flashings and built-up bituminous roofing. These guides may not be available new, however, since GSA now uses a modified version of AIA Service Corporation's *Masterspec* as its specification guide. Therefore, most pertinent requirements in the old GSA guides have been incorporated into this book.

For detailed wood shake and shingle requirements, contact the Red Cedar Shingle and Handsplit Shake Bureau. Request copies of *Certi-Split Manual of Handsplit Red Cedar Shakes* and other current data, including the latest recommendations for applying new wood shingles and shakes over existing materials. Much of the wood shingle and shake data included in *The NRCA Steep Roofing Manual* and the 1981 *Architectural Graphic Standards* is based on Red Cedar Shingle and Handsplit Shake Bureau recommendations.

John Harvey, in his 1972 book *Conservation of Buildings*, talks a little about stone roofing in England. For more detailed technical information, contact one of the historic preservation resources marked in the Appendix with a HP. Start with the National Trust for Historic Preservation.

In the event one encounters a reed-thatched roof in need of repair, which is highly unlikely in the United States today except on a rare historic preservation project, Donald W. Insall's 1972 book *The Care of Old Buildings Today: A Practical Guide,* contains some comments about rethatching work.

Low-Slope Roofing

Low-slope roofs are used on both residential and commercial buildings, but in most of the country, they are more likely to appear on large residential projects rather than houses and on commercial buildings. In warmer zones, however, low-slope roofing is also used extensively on houses. This chapter discusses both built-up and single-ply low-slope roofing.

Metal roofing and metal flashing roof accessories and specialties for low-slope roofing are discussed in Chapter 8.

Insulation for low-slope roofing and sprayed-in-place polyurethane foam roofing are discussed in Chapter 5.

Steep roofing is discussed in Chapter 6.

We tend to think of low-slope roofing as applicable to roof slopes of one inch per foot or less. Some single-ply roofing types, however, function quite well on much steeper slopes and on odd-shaped roofs and domes.

Built-up bituminous membrane roofing has been used since about 1860. Single-ply membranes did not make their appearance for another 100 years but quickly became a strong force in the roofing world. In 1988, single-ply roofing accounted for about half the overall roofing market and was used for almost 75 percent of roofing replacement and re-covering projects.

Requirements Common to All Low-Slope Roofing

The Roofing System

Low-slope roofing is a system of mutually dependent components selected to best solve the problem at hand. The system's expected life should be in line with the life of the building itself. The system must make economic sense relative to the cost of the building as a whole. The system must be the right kind for the roof deck. The system should be the best possible for the climate of the project site and should be constructed of readily available components.

Proper functioning of the roofing system is dependent on the supporting structure and roof drainage system. The roof structure must be able to support the loads that will occur. The roof slope must be sufficient to prevent ponding of water on the roof. There must be enough correctly located roof drains to drain the roof.

Roofing system components should be compatible with each other and surrounding materials. Each component has a unique function. Possible components include vapor retarder, underlayment, insulation, waterproof membrane, and ballast. Not all components are used in every system.

Most low-slope roofs are insulated. In a "compact" system, membrane, insulation, and vapor retarder, if there is one (see Chapter 2), are placed in contact (see fig. 2-3). In a "framed" system, the insulation is separated from the membrane by the structural deck or sheathing (see fig. 2-2). Where a ventilated air space occurs between the membrane and the insulation, framed systems are sometimes called "ventilated" systems. A system with two layers of insulation, one above the deck and another below, is a combination framed and compact system.

Here we are primarily interested in either uninsulated or compact systems. Insulation that is separated from the membrane may contribute to roof failure due to condensation but does not affect membrane installation or repair. Refer to Chapter 2 for a discussion of failures due to condensation.

The conventional order for the components in a compact system (see fig. 2-3) is, starting from the roof deck, insulation, vapor retarder (if there is one), insulation, membrane, and ballast. In a protected membrane roof (PMR system) the order is deck, underlayment, membrane, insulation, pervious fabric, and ballast (see fig. 2-4). The underlayment is needed to bridge the flutes in metal decking and might not be needed over a monolithic deck, such as concrete. A vapor retarder is not needed, because the membrane, being beneath the insulation, serves that function. With some insulation types, underlayment is needed in both systems when a fire rating is required.

Roof Decks and Underlayments

The roof deck must be sound, smooth, and properly designed to support the loads that will occur. Decks are classified as nailable, nonnailable, and insulated.

Nailable roof decks for low-slope roofs include wood planks, plywood, cement-wood fiber planks, lightweight insulating concrete, poured-in-place gypsum, and precast gypsum planks.

Nonnailable roof decks for low-slope roofs include precast and cast-in-place concrete.

Insulated roof decks for low-slope roofs include those covered with board insulation, sprayed-in-place polyurethane foam, or thermosetting insulating fill. Steel decks are considered insulated even when the covering board is actually a fire retardant or underlayment board installed solely to permit the membrane to bridge the deck's flutes.

Roof decks should be level or in plane, and open joints should be stripped or filled to prevent bitumen from dripping through them. Some precast units may require a topping to make them smooth. Thin materials, such as plywood, should have the joints supported to prevent deflection during roofing.

Concrete decks should be primed to absorb dust and provide a film to which bitumen or adhesive will adhere. Priming may not be necessary when the roofing is applied using coal-tar bitumen.

Nailable decks, where bitumen dripping is a possibility, should be covered by a mechanically fastened base sheet before built-up or single-ply roofing is applied.

Flashing

Flashings, which seal the gaps between roofing and other construction, are one of the most vulnerable parts of any roofing system. Many roof leaks are actually failed flashing problems.

Flashings should be compatible with the materials flashed. They should not cause physical damage, hasten deterioration, or differ radically in coefficient of expansion and contraction, and they should be capable of being bonded readily to the material being flashed.

Bitumen

Temperature. Temperature affects the flow rate of a bitumen. Temperatures that are too high make the bitumen too liquid and tend to decrease the amount of bitumen in the system, because the material flows too readily. Bitumen that has been heated to too high a temperature also tends to separate and leaves voids. Voids are potential leak points, especially when they stack up in a membrane system.

Temperatures that are too low produce bitumen that is not fluid enough, which will not adhere properly because it does not fuse with the bitumen in the felts. In addition, underheated bitumen has higher expansion rates and low tensile strength. All those problems can lead to roof membrane splits.

The optimum temperature range for bitumen is called its equiviscous temperature (EVT).

Every bitumen should be applied at its EVT and should never be heated above its flash point (FP) or any temperature noted by its producer as being a limit. In addition, asphalt bitumen should not be kept above its finished blowing temperature (FBT) for more than 4 hours. FBT is the temperature at which blowing of the asphalt was completed. Asphalt bitumen is manufactured by blowing air or oxygen through the heated (above 500° F) residue left after crude oil is distilled. The longer the blowing continues, the greater the bitumen's viscosity. EVT, FP, and FBT should be marked on bitumen packages.

Materials.   The bitumen used in bituminous low-slope roofing systems is either asphalt or coal-tar.

Asphalt bitumen is classified in accordance with ASTM D 312 in four types, as follows:

Type I: Dead level asphalt

Type II: Flat asphalt

Type III: Steep asphalt

Type IV: Special steep asphalts

Coal-tar bitumen is classified in accordance with ASTM D 450 in three types, as follows:

Type I: Coal-tar pitch

Type II: Waterproofing pitch

Type III: Coal-tar bitumen

Each asphalt and coal-tar bitumen type has its own EVT. The lowest type number has the lowest EVT (type I is lower than type II, and so on).

Dead level asphalt and types I and III coal-tar bitumens are limited to use on roof slopes of 1/2 inch per foot and lower. Flat asphalt can be used on slopes up to 1-1/2 inches per foot; type III (steep) asphalt is usable on slopes of up to 3 inches per foot; and type IV asphalt will work on slopes as high as 6 inches per foot. Type II coal-tar pitch is used only in waterproofing applications (see Chapter 4).

It seems logical that if steep and special steep asphalt bitumen is usable on such a wide range of slopes that it would make sense to use them everywhere. Such is not the case, however. Steep asphalts are less durable and harder to apply than are the lower viscosity materials, which means that not only is greater skill needed to apply them, but application errors are more serious. In addition, greater temperature latitude is possible without harm in the less viscous types I and II asphalts and in coal-tar bitumens.

The better method is to use each bitumen material on its rated slope, using steep asphalts only where they are necessary. Exceptions might be made in extremely hot or cold climates.

Both coal-tar and asphalt are excellent roofing bitumens. Which is the better, though, is somewhat controversial. Most people agree that coal-tar is inappropriate on slopes exceeding 1/2 inch per foot and most effective on slopes of 1/4 inch per foot. There is no agreement, however, about using coal-tar on slopes between 1/4 inch and 1/2 inch per foot. In any case, existing low-slope roofing may have either asphalt or coal-tar bitumen but seldom both, because they are incompatible. (Coal-tar bitumen is not necessarily incompatible with hard asphalt in felt coatings. Consequently, some roofing systems have coal-tar bitumen and asphalt-impregnated felt. Caution should be exercised when using such a system, however, to ensure that the bitumen and the felt coatings are compatible.)

Accessory Materials

Plastic Roofing Cement (Roofing Mastic) and Flashing Cement. These are made from cutback asphalt or coal-tar and solvents. Both are trowel-applied, and both contain reinforcing fibers, but they are not interchangeable. Flashing cement is the stiffer of the two, because it has more and longer fibers. It also comes with a range of solvent contents, to make it usable at various temperatures. Flashing cement should be used for vertical applications and on cants. Roofing cement is suitable for horizontal applications, such as at gravel stop flanges, and for general roofing repairs.

Cold-Applied Membranes and Coatings. These materials are applied without heating. They are used for patching membrane systems and for coating smooth-surfaced membranes.

Building Paper. Rosin-sized sheathing paper, heavy felt kraft paper, or reinforced sisal paper.

Fibered and Nonfibered Aluminum Coatings. Reflective coatings are sometimes used on smooth surfaced roofing and on exposed bituminous flashings to help keep the roofing and flashings cool and protect bitumens from deterioration due to exposure to sunlight.

Resaturants. Asphalt or coal-tar cutback liquid products used· to treat flood coats that have developed alligatoring or pinholes and felts that have dried out. These materials are self-sealing and will not alligator or form pinholes.

Walkway Protection Boards. Various products are used for walkway surfacing. They include asphalt-impregnated boards, mineral-topped insulation panels, and even concrete pads. There are also some PVC and polyester fleece-backed materials available for use as walkway surfacing today.

Warranties

The National Roofing Contractors Association's *Roofing Materials Guide* contains a detailed discussion of roofing warranties and compares the standard warranties of a number of roofing manufacturers.

Alan B. Stover's 1987 article ''A Specifiers Guide to Construction Warranties'' is a comprehensive discussion of construction warranties in general.

An existing roof may have a construction warranty that is still effective. The warranty might cover some of the damage being experienced when roofing fails. The chances are good, however, that if a warranty covers any part of the damage, it will be a small part with respect to cost. The existence of a warranty should be determined, and whatever help can be obtained as a result of it should be used. Expecting a warranty to cover the cost of repairing roofing damage, however, may lead to disappointment.

Built-Up Bituminous Membrane Roofing

Systems

Built-up bituminous membrane roofing is comprised of multiple layers of bitumen separated by reinforcing layers of felt. In an insulated built-up membrane system, the membrane covers, or lies beneath, one or more layers of thermal insulation (see Chapter 5). The top layer of the membrane is either a finished felt protected by a layer of bitumen or other coating material, or is covered and protected by a layer of ballast in a flood coat of bitumen.

In built-up bituminous roofing, the interlayer bitumen melts the bitumen in the membranes and fuses with it, welding the underlying and covering felt layers together. A complete built-up roofing system, therefore, acts as a unit. Because of this welding process, bitumen temperature is critical.

The number of felt and bitumen layers affect a roof's life. Each ply is usually considered worth 5 years of life. A four-ply roofing system, for example, is called a 20-year roof.

Built-up roofing is applied either in a conventional system where the membrane covers the insulation or in a protected membrane roof (PMR) where the insulation covers the membrane.

In most areas of the country, built-up roofing is usually covered with stone- or unit-paver-type ballast. Stone ballast is limited to slopes less than 3 inches per foot. Roof slopes for other types of ballast may also be limited. Mineral-surfaced roll roofing, which is usually not protected by ballast, is discussed in Chapter 6.

In some areas of the country, mineral-surfaced felt is often used as the top ply in built-up roofing systems, and no ballast is applied.

In built-up roofing systems on roofs where using aggregate ballast is impracticable, the top sheet is sometimes coated with bitumen or another coating material, such as fibered or nonfibered aluminum, in lieu of aggregate surfacing.

Aggregate-surfaced roofing is considered water-resistant. Both mineral-surfaced and smooth-surfaced roofing systems, however, are water-shedding rather than water-resistant. Organic felts are not appropriate for mineral-surfaced or smooth-surfaced roofing because of their tendency to absorb water. Even short-term ponding of water should be prohibited under all circumstances on mineral-surfaced and smooth-surfaced roofing.

For additional discussion, refer to "Requirements Common to All Low-Slope Roofing," given earlier in this chapter.

Materials and Their Uses

Built-up roofing materials include vapor retarder; insulation; asphalt or coal-tar bitumen; felts; and ballast. Vapor retarders are discussed in Chapter 3. Insulation is discussed in Chapter 5.

Polymer-modified bitumen roofing materials, which are also used in built-up roofing systems, are excluded here. They are discussed later in this chapter under "Single-Ply Membrane Roofing."

For additional materials used in built-up roofing systems, refer to "Requirements Common to All Low-Slope Roofing," discussed earlier in this chapter.

Felts.   Built-up bituminous roofing felts are nonwoven mats of organic or inorganic fibers bonded together and saturated with asphalt, modified asphalt, or coal-tar bitumens. Felts are either fully saturated (saturated), partly coated (impregnated), saturated and coated with asphalt which has been stabilized with a finely ground mineral (coated), or saturated and coated with a ground mineral, such as slate or colored rock (mineral-surfaced). Saturated felts and impregnated felts are used as ply or base sheets. Coated sheets are used as base sheets. Mineral-surfaced felts are used as top (cap) sheets, exposed base flashing, or roll roofing (see Chapter 6). Sheets for roll roofing may be fully coated, coated except for a 2-inch selvage, or designed such that roughly one half is surfaced and the other half is unsurfaced. Grooved mineral surface felts are sometimes used as vented base sheets.

Felts are designated by a number representing their nominal single-ply weight per square (100 square feet) of roofing. Thus, the nominal weight of number 15 felt is about 15 pounds per square of felt. The actual felt weight is a little less than the nominal.

Felt Fibers: Cellulose, Asbestos, and Glass. Cellulose fibers are used in organic felts. These are the least costly felts.

Asbestos fibers were used in some felts in the past and may still be found in existing roofs. Some felts containing asbestos fibers also contained organic fibers.

Glass fibers are bound together into a mat using a thermosetting binder, phenol-formaldehyde, or urea-formaldehyde.

Bitumen. Both asphalt and coal-tar bitumens are used in coating, impregnating, and saturating felts.

Surfacing of Felts. Felts are manufactured either smooth-surfaced, coated, or mineral-surfaced. They may be surfaced after installation with protective or reflective coatings.

Materials in the coating bitumen used in manufacturing coated felts include talc, silica, slate dust, dolomite, trap rock, and mica.

Materials used in coating mineral-surfaced roofing felts include ceramic, slate, and colored rock granules. Mineral-surfaced sheets may be either fully or partially coated.

Prepared roofing is usually dusted on the back side with talc or another finely ground mineral to prevent felts from sticking together when rolled.

Modified Bitumen Felts for Use in Making Repairs. Same materials as discussed under "Single-Ply Roofing" later in this chapter.

Ballast. Ballast resists membrane, or insulation in PMR, lifting by the wind and protects the bitumens, or insulation, against solar radiation, air pollution, and impact damage which can result from people walking on the roof or from hailstones falling on it.

Many materials are used as ballast in built-up roofing systems. Among them are stone and slag aggregate, concrete pads, and many different types of unit masonry and tile. When the ballast is pavers that will be used routinely by pedestrians or vehicular traffic, the roofing should probably be classified as waterproofing, which has some characteristics that are different from roofing (see Chapter 4).

The best aggregate ballast is either washed river gravel, crushed stone, or a fused by-product of iron production called blast-furnace slag. Of the three, slag is usually considered superior. Many other materials have also been used as surfacing aggregate, but some have produced inferior installations and reduced membrane life.

Aggregate ballast is embedded in a heavy flood coat of bitumen. This flood coat adds another waterproofing layer to the system, which is a major advantage of aggregate ballast over some other ballast types. It is possible to make the layer of bitumen surrounding the ballast heavier than the layer coating a smooth-surfaced roof. Unfortunately, the ballast makes visual inspections of the membrane impossible, and repairing, re-covering, or replacing the roofing both difficult and expensive.

Aggregate ballast size, nominally 3/8 inch in diameter, is critical. Fines in undersized aggregate sink into the flood coat and interrupt the waterproofing layer, which makes the roofing more prone to leaks and decreases roofing life. Individual pieces in undersized aggregate are less likely to be bonded to the bitumen. Oversizing is also a problem, because the large voids between the stones can trap foreign materials, but is less likely to occur.

Some roofs have two layers of top coating and aggregate surfacing, which is expensive but seems to extend roofing life.

Installation

There are many built-up bituminous roofing systems used today. Each membrane manufacturer has a variety of different systems. Number of plies, rates of bitumen application, felt types, underlayments, and surfacing differ from system to system and manufacturer to manufacturer. There are some basic criteria that follow through all systems, however.

Plies should be laid shingle fashion starting at low points so that water will flow across the system without interruption.

Most systems have at least four plies. On nailable decks, the membrane roofing should be placed over a layer of rosin-sized building paper. Plies should be installed in accordance with the roofing manufacturer's specifications for the system selected. The National Roofing Contractors Association's 1986 *The NRCA Roofing and Waterproofing Manual* includes detailed generic specifications for most roofing systems, including those installed over existing materials.

Flashing for Built-Up Roofing.   Cap flashings and some penetration and expansion joint flashings are metal (see Chapter 8). In some, particularly older, roofs, base flashings may be metal. But base flashings and stripping felts at metal flashings and roof specialties, such as gravel stops, should be bituminous. Some penetration and expansion joint flashings are elastomeric.

Sheet materials used for flashing (other than metal) include plied felt, fabric-based sheets, composite bituminous sheets, and plastic sheets. Roofing and flashing cements are also a major component of flashings (see "Accessory Materials," found earlier in this chapter). The term "composition flashing" refers to any bituminous flashing. Composition flashing may be either fabric or felt. The flashing used is a component of the selected system. Manufacturers' specifications usually specifically state which flashing materials should be used. NRCA's 1986 *The NRCA Roofing and Waterproofing Manual* also states the type of flashing that should be used in each roofing system it discusses.

Plastic flashings include vinyl, neoprene, butyl rubber, and others. Some plastic flashings are not compatible with asphalt and should not be used in asphalt built-up roofing systems. Compatibility should be verified with the materials manufacturers in every case.

Composition flashings and plastic flashings are applied in built-up roofing systems using hot-mopped asphalt or cold-process bituminous flashing cement. Because hot-mopped flashings should be nailed to the adjacent vertical surfaces, they are not appropriate where the surfaces are not nailable. Steep asphalt is sometimes used in lieu of asphalt flashing cement.

Where the roofing system or roof flashing details require, flashings should be securely fastened in place. Flashings extending up vertical surfaces should always be fastened at not more than 8 inches on center. Where the vertical surface is subject to differential movement relative to the roof, an expansion joint should be imposed (see Fig. 7-1). A wood curb should be attached to the roof structure and free of the wall. The base flashing should then extend up the curb and be free of contact with the wall. Metal cap flashing should bridge the expansion joint and be fastened to the wall, not to the curb or the base flashing. Refer to "Where to Get More Information" in Chapter 8 for a listing of sources of roof flashing details.

Nonmetal flashings are often coated with fibered or nonfibered aluminum material. Aggregate-surfaced felts used as base flashing are not usually coated. Aluminum-foil-faced felts may also be used as top felts in flashings without a coating.

Base Flashing.   Sheet metal should not be used as base flashing for built-up roofing. Base flashing is usually inorganic composition flashing, because it has the same expansion coefficient as the roofing, and good weather and corrosion resistance. Aggregate-surfaced sheets are often used.

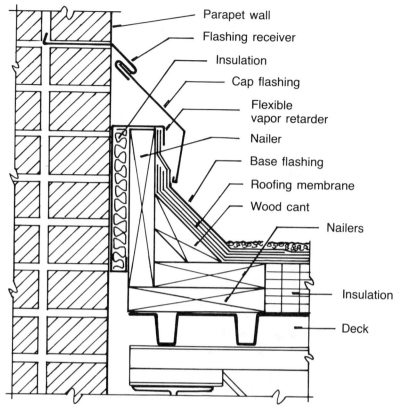

Parapet wall
Flashing receiver
Insulation
Cap flashing
Flexible
vapor retarder
Nailer
Base flashing
Roofing membrane
Wood cant
Nailers
Insulation
Deck

Figure 7-1 Roof edge expansion joint for use where differential movement is likely between roof deck and wall.

Base flashing should have the same number of plies as the roofing. Vertical legs should be mopped to primed vertical surfaces. The actual construction, including number of plies, materials, and application methods, of the base flashing should be in accordance with the system selected.

Provision should be made to permit differential movement between base and counter (cap) flashing. They should not be locked, fastened, or sealed together.

Roof Drains. Here the prohibition against metal flashings breaks down. Roof drain flashings are almost always metal. Refer to Chapter 8 for a discussion of metal roof drain flashings.

Roof Penetration. Large penetrations should be curbed and flashed using composition base flashings. Small penetrations are usually flashed using sheet metal (see Chapter 8). Sometimes, formed plastic materials have been used as penetration flashings in built-up roofing.

Roof Specialty Flashing.   Roof specialties, such as gravel stops, should be stripped into the roofing using strips of composition flashing material set in bituminous roofing cement.

Single-Ply Membrane Roofing

Systems

In a compact single-ply membrane system, the membrane covers (conventional) or lies beneath (PMR) one or more layers of thermal insulation (see Chapter 5). In a framed system, the single-ply membrane is usually applied directly over the structural deck. In a system where the membrane is exposed to the weather, the top surface of the membrane is either prefinished, field-coated, or covered and protected by a layer of ballast. In a protected membrane roof (PMR), the insulation is held in place by aggregate or other ballast.

Single-ply roofing membranes are either loose laid, fully adhered, partially adhered, or mechanically fastened in place.

Each singled-ply membrane manufacturer has specific requirements for the roofing systems applicable to its product. Those recommendations must be followed if a warranty is to be granted, and unless obviously incorrect, they should take precedence over other recommendations. Codes and insurance requirements, of course, always have the final word.

Polymer-modified bitumen roofing membranes are used both as single-ply roofing and as the membrane in built-up roofing systems. Their proper category is yet another in a seemingly endless list of roofing industry controversies. Including modified bitumens in the single-ply roofing category should not be construed to mean that they are properly so labeled. The selection was purely arbitrary.

Materials and Their Uses

Single-ply roofing materials include vapor retarder, insulation, bitumen and adhesives, membrane, and ballast. Vapor retarders are discussed in Chapter 3. Insulation is discussed in Chapter 5.

For additional materials used in single-ply roofing systems, refer to "Requirements Common to All Low-Slope Roofing," listed earlier in this chapter.

Membrane Sheets. There are many single-ply roofing membrane sheet types available. Even though an attempt has been made to include the major types in use at the time of this writing, the following list is only partial. Types seldom used in 1988 may be more popular now. New products are being introduced regularly. By now there are certainly products available that are not mentioned here. There may even be entire new classes of products.

The materials listed are either factory-cured (vulcanized) elastomers, uncured (nonvulcanized) elastomers, thermoplastics, or polymer-modified bitumen. Most uncured elastomers cure in place after application, after which they are said to be "thermoset." An uncured elastomer can be molded, shaped, and welded by heat. A vulcanized or thermoset elastomer cannot be molded, shaped, or welded by heat. Thermoplastics never cure, which means that they can be softened and reshaped by application of heat throughout their life. The designation of materials in the list as vulcanized elastomeric, uncured elastomeric, or thermoplastic was determined by review of manufacturers' statements about their products and other industry sources. When sources did not agree, an arbitrary choice was made. Sometimes, manufactureres use the terms "elastomeric" and "thermoplastic" interchangeably, even though they have different meanings. Some manufacturers say that their elastomers are thermoplastics until they cure, which may be true, but such an assertion tends to lead to confusion.

The sheet thicknesses mentioned are standard today. There is, however, a trend toward thicker membranes.

EPDM (Ethylene Propylene Diene Monomer). EPDM is an elastomeric compound made from the three materials which make up its name. EPDM sheets are generally only 30 to 35 percent EPDM. The remaining materials are plasticizers, accelerators, and fillers, such as carbon black. EPDM sheets are usually either black or white. They vary in thickness from 30 to 60 mils. Loose laid membranes are usually 45 mils thick. Adhered membrane is usually 60 mils thick. Most EPDM is vulcanized.

Neoprene (Polychloroprene). Neoprene is a vulcanized synthetic rubber sheet material, which is usually 60 mils thick for roofing and which may be reinforced. Neoprene is also available nonvulcanized for use as flashing. Some formulations will accept a liquid CSPE color coating.

CSPE (DuPont Hypalon). CSPE is nonvulcanized elastomeric laminate consisting of a chlorosulfonated polyethylene weathering layer and a mineral-fiber backing, and, in some products, a polyester scrim reinforcing layer. Sheets are usually 60 mils thick. Neoprene is also available in liquid form for use as a coating. Sheet and liquid neoprene are produced in many colors.

CPE (Chlorinated Polyethylene). CPE is a nonvulcanized elastomeric sheet reinforced with polyester, which is usually 40 to 48 mils thick and can be pigmented with many colors.

PIB (Polyisobutylene). PIB is a nonvulcanized elastomeric sheet formulated from elastomer, carbon black, and other additives, which is a 60-mil-thick sheet laminated to a 40-mil polyester backing.

PVC (Polyvinyl Chloride). PVC is a thermoplastic sheeting with plasticizers, from 40 to 60 mils thick. PVC is available both reinforced with glass fiber or polyester and not reinforced. It comes without finish for field finishing with coatings and with factory-applied acrylic coatings for weather exposure.

NBP (Nitrile Alloy). NBP is a thermoplastic sheet composed of a polyester fabric coated with a compound made from butadiene-acrylonitrile copolymers, plasticizers, and other additives. It is 30 to 40 mils thick.

CPA (Copolymer Alloy). CPA is a thermoplastic blend sheeting, with many ingredients, one of which is PVC. It is reinforced with polyester fabric and formed into sheets that are about 32 mils thick.

EIP (Ethylene Interpolymer). EIP is a thermoplastic blend sheeting, with many ingredients, one of which is PVC. It is reinforced with polyester fabric and formed into sheets that are about 32 mils thick.

Polymer-Modified Bitumen. Polymer-modified bitumen is a sheeting made from asphalt modified with atactic polypropylene (APP), styrene-butadiene-styrene (SBS), or styrene-butadiene-rubber (SBR), and reinforced with glass fiber or polyester mats or scrim, polyethylene sheets, or aluminum, copper, or stainless steel foil. APP-modified sheets *must* be reinforced. Some polymer-modified bitumen sheets are finished at the factory for weather resistance and appearance using black, white, or colored granules. Others are suitable for field application of coatings or granules for weather resistance and appearance. APP-modified sheets *must* be coated with granules. Polymer-modified bitumen sheets are from 40 to 60 mils thick.

APP-modified bitumen sheets are suitable for torch application. They are not compatible with asphalt. SBS- and SBR-modified bitumen sheets are compatible with asphalt and may be used in built-up roofing systems. They may also be torch-applied or self-adhering.

APP-modified sheets are better for warm weather applications. SBS- and SBR-modified sheets are better in cold weather applications.

Fasteners.   Many types of mechanical fasteners are used to install single-ply roofing. They vary with roofing application methods, roofing membrane materials, type of deck, and type and thickness of insulation. Fasteners should be noncorrosive.

Ballast.   Most ballast used in single-ply roofing systems is washed river gravel. Concrete pads are also sometimes used.

The recommended aggregate ballast for modified bitumen roofing is the same as that discussed earlier in this chapter for built-up roofing.

Adhesives and Solvents.   Different manufacturers' recommendations vary for their systems. Adhesives and solvents should be the materials recommended by the roofing system membrane manufacturer. Adhesives glue the materials together but do not weld them. Commonly used adhesives include one-part-butyl, two-part-butyl, and neoprene. Solvents weld single-ply roofing materials together.

Installation

Single-ply roofing membrane on existing roofs may have been laid loose, adhered to the deck or insulation using hot- or cold-applied bitumen or adhesives, self-adhered to the substrates, or mechanically fastened in place using a variety of fastener types.

Polymer-modified asphalt roofing materials may be used in single-ply roofing, as discussed below, or in built-up roofing systems. In built-up roofing applications, they are subject to the same requirements applicable to built-up bituminous systems constructed using other membranes, as discussed in the portion of this chapter entitled "Built-Up Bituminous Membrane Roofing."

Refer to Chapter 5 for a discussion about whether the insulation in a single-ply roofing system should be fastened in place or laid loose with no fasteners.

There may be restrictions on the type of insulation that should be used with a particular membrane, especially in adhered systems. The manufacturer's recommendations should be followed.

Recommendations for the number and location of mechanical fasteners and the type, amount, and placing of bitumen or adhesive are offered by manufacturers, industry member associations, such as the National Roofing Contractors Association, Factory Mutual (FM), Underwriters Laboratories (UL), and guide specifications such as *Masterspec* and those produced by the Naval Facilities Engineer Command. Unfortunately, the recommendations do not always agree. Some sources refer to each other for requirements. See the section on "Where to Get More Information."

Plies should be laid shingle fashion starting at low points so that water will flow across the system without interruption.

Loose-laid and mechanically attached single-ply membrane roofing should be placed over a layer of rosin-sized building paper, unless the membrane manufacturer objects. Some manufacturers may recommend a layer of another separator material, such as insulation board, instead of building paper. The membrane should be installed in accordance with the roofing manufacturer's specifications for the system selected. NRCA's 1986 *The NRCA Roofing and Waterproofing Manual* includes detailed generic specifications for most roofing systems, including those installed over existing materials.

Loose-Laid Roofing   Loose-laid membrane is placed over the substrates with only minimal fastening at the edges and around penetrations and held in place by ballast.

Loose-laid single-ply roofing is susceptible to wind uplift damage and may be pulled away from flashings at the roof edges because of cumulative thermal changes in the roofing membrane. Loose-laid roofing, therefore, must be held down by ballast. Besides resisting membrane lifting by the wind, ballast protects the membrane against solar radiation, air pollution, and impact damage.

The Single Ply Roofing Institute's publication (SPRI) *Wind Design Guide for Ballasted Single Ply Roofing Systems* offers guidance for analyzing wind uplift.

When concrete pads are used as ballast, care must be taken to prevent damage to the membrane by the pads. In addition, concrete pads must be carefully aligned, or slotted pavers must be used, to prevent the pad from blocking water flow over the roof.

Fully Adhered Roofing.   In fully adhered systems, a single-ply roofing membrane is completely attached to the substrates using hot- or cold-applied bitumen, cold-applied adhesives, or solvents by heating (torching) the back of the membrane or by pressing self-adhering membrane in place.

The membrane manufacturer's recommendations should be followed for the type of adhesive, solvent, or bitumen, application rates, and installation methods.

Partially Adhered Roofing.   There are two ways to install partially adhered single-ply roofing. In one method, a bitumen, adhesive, or solvent is applied in beads or strips covering about one half the substrate and the membrane is rolled into the adhesive or solvent.

In the second method, mechanical fasteners with top plates are driven into the deck, an adhesive or solvent is placed on the plates, and the membrane is pressed into the adhesive or solvent. Sometimes, but not always, the fasteners are the same ones used to anchor the insulation.

The membrane manufacturer's recommendations should be followed for the type of bitumen, adhesive, or solvent; application rates; size and location of adhesive strips or beads; and installation methods.

Mechanically Fastened Roofing.   Mechanically fastened single-ply roofing is attached using either a penetrating or a nonpenetrating method. For the penetrating method, fasteners are driven through the membrane, sometimes through a metal or rubber bar, and covered with a layer of membrane.

For the nonpenetrating method, fasteners are anchored to the structural deck and the membrane is clamped to the fastener. The fastener does not penetrate the membrane.

Joints.   Some single-ply membranes have a talc or mica coating to keep them from sticking together when rolled. The coating must be removed with a solvent before joints are made. The solvent material and cleaning method varies with the membrane material.

Joints in nonvulcanized elastomeric materials are heat- or solvent-welded, which fuses the two sheets together in a joint sometimes stronger than the welded sheets. Joints in vulcanized or field-cured (thermoset) elastomeric materials are sealed using adhesives. Some materials require contact adhesives, which must be applied to both layers of the material being joined. Some materials can be sealed with generic adhesives. Others require adhesives made from the same compounds as the sheets. Some materials are naturally self-adhering. Others have been made self-adhering by an adhesive coating applied over the entire sheet. Others have self-sealing strips bonded to the sheets along the edges. Review the manufacturer's specifications to see which joint sealing method is appropriate for the product selected. Do not assume that all products of a single material are sealed alike. Neoprene, for example, is available both vulcanized and nonvulcanized.

Joints in thermoplastic materials such as PVC or EIP may be either heat- or solvent-sealed.

Joints in self-adhering modified bitumen sheets are made by pressing the sheets together and rolling. Joints in nonself-adhering SBS- or SBR-modified bitumen sheets are either torch-welded or sealed with hot asphalt. Joints in nonself-adhering APP-modified bitumen sheets are torch-welded.

Flashing for Single-Ply Roofing.   Base flashings and stripping over penetration flashings, roof drain flashings, and roof specialties, such as gravel stops, are usually the same material as the roofing membrane or a nonvulcanized compatible elastomeric sheet. Uncured neoprene (CSPE), for example, is sometimes used to flash EPDM. Metal base flashings are not generally used. Provision should be made to permit differential movement between base and counter (cap) flashing. They should not be locked, fastened, or sealed together.

Most cap flashings and some penetration, expansion joint, and roof drain flashings and roofing specialties, such as gravel stops, are metal (see Chapter 8). Sometimes the metal is coated or clad with the same material used in the membrane. The coating or cladding makes sealing joints between flashing or roofing specialty and membrane easier and more positive. Some penetration, expansion joint, and roof drain flashings, and some roofing specialties are formed or molded from plastic materials, such as neoprene, vinyl, and butyl rubber. Manufacturers usually state which flashing and specialty materials are appropriate. NRCA's 1986 *The NRCA Roofing and Waterproofing Manual* also recommends the type of materials that should be used in each roofing system it discusses.

Joints between flashings and roofing membranes are sealed the same as joints in the membrane.

Where the roofing system or flashing details require, flashings should be securely fastened in place. Flashings extending up vertical surfaces should always be fastened at not more than 8 inches on center. Where the vertical surface is subject to differential movement relative to the roof, an expansion joint should be imposed (see Fig. 7-1). A wood curb should be attached to the roof structure and be kept free of the wall. The base flashing should then extend up the curb and be free of contact with the wall. Metal cap flashing should bridge the expansion joint and be fastened to the wall, but not fastened to the curb or the base flashing. Refer to "Where to Get More Information" in Chapter 8 for a listing of sources of roof flashing details.

Coatings and Surfacings.   Acrylic emulsion coatings are used to cover PVC and some modified bitumen membranes for improvement of the appearance. Some modified bitumen membranes may require application of mineral granules in the acrylic emulsion coating or aggregate in a bituminous flood coat for fire resistance. Fibered and nonfibered aluminum coatings are also used on some modified bitumen membranes. Liquid Hypalon is sometimes used to coat EPDM and some formulations of neoprene to provide a color coating. Silica sand coatings are sometimes used on EPDM.

Low-Slope Roofing Failures and What to Do about Them

Low-slope roofing fails because of poor roofing system design, improper installation, bad materials, natural deterioration due to age, failure to maintain the roof, and leaks through flashings or adjacent materials.

Refer to Chapter 5 for discussions of the effect of roofing membrane, flashing, and adjacent construction leaks on insulation, of the controversy about using roof vents in board insulation systems, and of the necessity for and methods of ventilating poured-in-place lightweight insulating concrete.

Reasons for Failure

All roofing membranes will eventually suffer natural deterioration and fail due to age. Most premature failure is due to poor design, improper installation, or the use of bad materials.

Poor Design.   From the first design sketches up to the last word written in the construction specifications, the designer makes decisions that will dictate success or failure of a building's roof. Bad design results from many causes, including decisions forced on the designer by the owner.

The owner's bad judgment can range from demanding a particular roofing type, which may be the wrong one for the case at hand, to insisting on using cheap roofing products to save money. The wrong roofing type can fail because of environmental or man-made factors. Materials that are sensitive to animal fats and vegetable or synthetic oils, for example, as some synthetics are, are a poor choice for roofing a commercial kitchen. Money saved by using cheap roofing is often spent many times over in repairs and premature re-covering or replacement.

The owner's bad judgment is sometimes exacerbated by the designer's failing to inform the owner of the consequences of decisions. Sometimes, the information is available, but designers do not take the time to inform themselves. Sometimes roofing manufacturers, contractors, or their associations either fail to recognize a potential harm or recognize a problem but do not inform the rest of the industry.

Most bad design decisions, however, are the designer's fault. Too often, designers lack the necessary knowledge to properly design roofing systems and are too lazy or unconcerned to learn them. Some of them are not even aware of their own incompetence. Others are so preoccupied with aesthetics that they ignore "mundane" things such as roofing. Some roof failures occur because the designer is simply too lazy to carry the design process to its conclusion, leaving detailing and specifications to unsupervised incompletely trained subordinates.

Good roof design starts with basic building design. Some roof shapes are almost impossible to roof successfully. Poor design can lead to failed roofs that are impossible to repair (see Fig. 7-2).

The structure must be stable, without excessive deflection, and sufficiently strong to support the roofing and all applied loads. The roof deck must also be strong enough to support the roofing without undue deflection. Some designers believe that roof structures and decks should be overdesigned to carry additional layers of roofing that will be applied later. When an existing roof eventually fails due to aging, new roofing, sometimes including new insulation, is often installed directly over the existing roofing (called re-covering) without removal of the existing materials. Failed substrates will cause roof failure even when the roofing is properly selected, detailed, and installed.

Roof deck material and configuration affect which roofing system can be used.

Figure 7-2 Sometimes, roofing problems defy repair. Then, as in the building in this photo, the only solution may be to build an entirely new roof structure over the original. (*Photo by author.*)

Roof size must be taken into account in good roof design. Large roofs, for example, are difficult to ventilate. Peripheral vents alone may not work. In addition, expansion and contraction control becomes important in large roofs. Expansion joints provided to accommodate structural movement are usually extended through the roofing. Too often, though, designers ignore roof expansion and contraction except where building expansion joints occur. Roofs less than 100 feet in any dimension seldom need expansion or contraction control. Roofs more than 300 feet long, however, should be divided into sections between 150 and 200 feet long. The method of dividing large roofs between structure expansion joints is important. Elastic control joints (sometimes called expansion joint covers) are not appropriate. The proper method is to build area dividers, which consist of a double wood member curb at least 8 inches high across the entire roof (see Fig. 7-3). The double curb is flashed with composition base flashing and capped with metal flashing as would be an expansion joint, but the two wood members are placed in contact and there is no structural interruption, so there is no expansion joint.

Probably the most important single factor influencing built-up roofing success or failure is roof slope. Designers seem reluctant to want to deal with the aesthetic problems caused by sloping roofs at the rates necessary

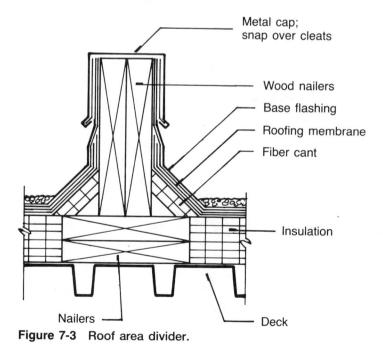

Figure 7-3 Roof area divider.

to prevent roof leaks. Designers want roofs to be flat, so that visible roof edge lines are straight and level. To drain properly, low-slope roofing must slope at least 1/4 inch and preferably 1/2 inch per foot. Roof slope is important in all roofs, but even more important in large roofs, because they are more likely to have low spots where water can pond.

Slope requirements for single-ply roofing vary from product to product and system to system. The Single Ply Roofing Institute (SPRI) recommends that aggregate-ballasted single-ply roofing not be used on roofs with slopes of more than 2 inches per foot. Some manufacturers insist that a sloped roof is not essential with single-ply roofing; other sources point out that roof slopes of 1/4 or 1/2 inch per foot help prevent ponding and buildup of pollutants which can damage the roofing, especially in roofs with aggregate coverings. Single-ply roofing has been used on very steep slopes, of course, but there are no standards, and little agreement, about slope limitations for the various applications systems.

Materials that give no problems when used alone can cause great difficulty when combined into a roofing system. Insulating a roof, for example, may cause condensation problems that must be dealt with. The solution is to introduce another roof component, a vapor retarder (see Chapter 2), which may require introduction of roof vents, some of which penetrate the roofing and require flashing, and so on. Leaving out a necessary component of a roofing system or using the wrong material as a component will lead to failure. The wrong material is one that is incompatible with the other components or inappropriate for the system.

Poor design also includes selecting inferior materials or systems. Some inferior material selection is done out of ignorance, but more often is done to save money. Poor quality will lead to failure no matter how the installation is made. Selecting inferior materials or systems to save money is false economy, actually resulting in more cost in the long run.

Bad detailing and poorly drafted specifications contribute to roof failures. Details are often drawn by inexperienced lightly supervised people who have little knowledge of roofing installation and are too busy or too lazy to learn it. Most improper detailing happens at perimeter and flashing conditions. Too few membrane plies are used and base flashings are not extended high enough up vertical surfaces. Cap flashings do not lap base flashings enough. Cant strips are omitted. Differential movement is ignored. Specifications errors include calling for the wrong bitumen or adhesive, not requiring protective coatings, either not specifying or incorrectly specifying fasteners, and specifying the wrong joint sealing materials or methods.

Often, even competent designers are frustrated when trying to design successful roofing systems, because the building industry has been slow to develop roofing standards. Some existing standards conflict. Knowledgeable people strongly disagree on some roofing related subjects. Codes do not agree with industry standards. Requirements for roof supports are not universally consistent with good roofing practice. For example, some standard span deflection tables for roof structural elements, such as bar joists, permit sufficient roof deflection to result in ponding of water, which can contribute to roofing failure.

Improper Installation.   As serious as poor design can be, almost everyone in the building industry agrees that most roofing problems are caused by poor workmanship during installation. Some poor workmanship occurs because the roofers are ill-trained or inexperienced and poorly supervised. Most, however, is a result of economic and time pressures. The damage shown in Figure 7-4 resulted from using improper installation techniques.

Some improper installation results from not looking at the design drawings and specifications or manufacturer's installation instructions. Some contractors look at those documents, assume them to be incompetent, and rely instead on their own experience. Unfortunately, new products sometimes make experience useless. A single-ply roofing sheet may look and feel like rubber but not behave like rubber in any way.

Figure 7-4   Incomplete adhesion in this joint due to improper installation technique. (*Photo courtesy of The Garland Co., Inc.*)

Improper installation includes using the wrong fasteners, over- or under-driving fasteners; not properly priming surfaces to receive bituminous products; using too little, or too much, bitumen or adhesives; leaving out flashing plies or using the wrong type of plies; leaving out, or leaving gaps in, cant strips; and not properly sealing seams.

Probably the worst problem in roofing, and the most likely to happen, is trapping moisture within the system during application. Roofing system components should be installed only during dry weather and should be kept dry continuously until the system has been completely installed.

Bad Materials.   Bad roofing materials sometimes arrive at a construction project, but the number of times this happens is not excessive. Delivery of the wrong materials is more of a problem.

Natural Degeneration and Failure to Maintain.   All roofing materials have finite life spans and will eventually fail due to aging and the effects of weather, air pollution, and wear and tear. The natural life spans of materials used in low-slope roofing can be drastically shortened if they are not properly maintained. Conversely, diligent maintenance can extend a roof's life far beyond its rated life.

Regular maintenance, such as recoating smooth surface roofing, covering aggregate ballast bare spots, and cleaning drains can have dramatic effects, often resulting in savings that greatly exceed their costs.

Evidence of Failure

Failed or damaged roofing and joints do leak, of course, but leaks so often result from flashing failure that unless roofing damage is obvious, it is appropriate to first look at the flashings when confronted with a roof leak. Refer to Chapter 8 for a discussion of potential metal flashing problems.

Some roof leaks result from substrate failure. When that happens, the structural failure must be corrected before any attempt is made to repair the roofing.

Some apparent roof leaks are due to condensation. Refer to Chapter 2 for a discussion of condensation problems.

Some apparent roofing problems are actually leaks through adjacent construction, such as masonry parapet walls. When the cause of a roof leak cannot be found in roofing or flashing, the adjacent structure should be suspected.

Some problems are due to fastener failure, often caused by water intrusion. The water rusts the fasteners or the deck around them or causes the insulation to exude acids which destroy the fasteners or the deck at the fasteners. Fasteners can also back out (see Fig. 7-5) because they were overdriven initially, stripping the threads. See Chapter 5 for additional discussion.

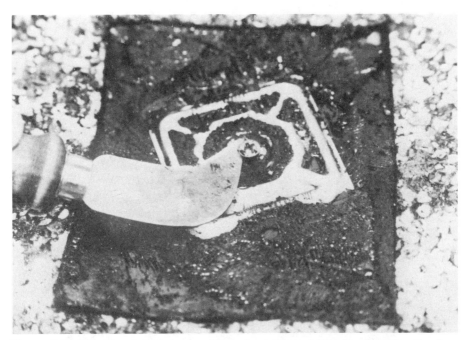

Figure 7-5   Fastener backing out. (*Photo courtesy of The Garland Co., Inc.*)

Failure can often be detected by inspection before leaks occur.

The first symptom of built-up roof membrane failure will probably be blisters, splitting, or ridging. Sometimes membranes will slip, delaminate, or alligator, or their surfaces will erode. Some of those defects will be caused by wet insulation, but wet insulation is caused by leaks, and so the circle closes.

Ballasted roofing system failures may also show up as missing aggregate ballast.

Loose-laid single-ply roofing will lift in the wind and displace part of the ballast. Repeated storms will cause the ballast that holds down the roof beneath the piles to congregate. The unballasted membrane will then raise even more between the ballast piles, eventually splitting the membrane.

The most dramatic roof failures are wind blow-offs (see Fig. 7-6). Roofs are not designed to withstand the very high winds associated with hurricanes and tornadoes, so blow-off during those types of storms is to be expected. Roofing blow-off by lower-speed winds can be prevented by properly designing the edge details and by fastening the roofing and insulation securely to the deck. Sometimes, parapets will help prevent membrane blow-off, but high parapets may make the situation worse instead of better.

Figure 7-6 Wind blow-off. (*Photo courtesy of The Garland Co., Inc.*)

Most blow-offs begin when the edge flashing or gravel stop and its nailer are peeled back by the wind, permitting wind uplift forces to drive beneath the membrane, insulation, or both and lift them from the roof.

Some wind uplift failures, especially in loose-laid, partially adhered, and mechanically fastened single-ply roofing, start with what is called "ballooning." The loose membrane lifts from the insulation or deck and undulates in a billowing wave. Forces in the membrane cause it to either split or pull away from the edges, resulting in a major blow-off.

Other factors that contribute to blow-off include inadequate, incorrectly located, or corroded fasteners, improperly installed or incorrectly selected adhesives, and excessive roof deck deflection.

A related problem is gravel blow-off, which occurs when the gravel is too small and loose.

Severe weather can cause other types of damage as well. Hailstones can make indentations in roofs where the insulation is not strong enough to resist the impact. Wind-driven hail can tear into a roof, dislodging the aggregate surfacing and ripping the membrane to shreds.

Some older membranes, especially PVC, will occasionally be found damaged by plasticizer migration, caused by membrane exposure to incompatible materials, such as asphalt, sunlight, or polystyrene insulation. Plasticizer migration may be evidenced by the membrane separating from flashings, becoming brittle, and cracking. Because of changes in formulation, thicker membranes, and reinforcement, modern membranes are less likely to suffer plasticizer migration problems in the short term. All plasticizers, however, will eventually leave the membrane.

Some membranes, especially white membranes, are subject to attack by algae and fungi, which turns them dark and causes the membrane surface to craze (see Fig. 7-7).

Inspection.   Every roof, especially roofs approaching their expected life spans, should be thoroughly inspected at least once a year. Some experts believe that low-slope roofing and flashings should be inspected twice a year, in the spring and fall. Repairs, metal flashings, and pitch pans, should definitely be inspected twice a year. Inspections should be made by persons who are knowledgeable about roofing and flashing problems. The purpose

Figure 7-7   Fungal growth on Hypalon roofing. (*Photo courtesy of The Garland Co., Inc.*)

of routine inspections is to detect problems early so that they can be repaired before major damage is done.

Roofing and composition or membrane flashing damage may be apparent in many ways before actual leaks occur, including the appearance of bare spots, breaks and separations in the felt, bubbles, soft spots, blisters, ridges, wrinkling, alligatoring, open joints, felt delamination, and separation of felts from substrates.

Refer to Chapter 5 for a discussion of some visual clues to wet insulation and a discussion of nondestructive testing for leaks and test cuts to verify the extent of wet insulation and membrane damage.

When a leak occurs, the most important thing to do is prevent further damage and stop up the leak or cover the leaking roofing with a temporary tarpaulin or plastic membrane.

After the leak is under control, look for the source of the leak. Examine the underside of the roof first by removing portions of the ceiling or entering the space above the ceiling. After locating the source from below, inspect the flashings and roofing from above to determine the actual source of the leak. Sometimes, multiple drips result from a single hole through a membrane. The best time to locate the source of a leak is while the water is coming in. It may be necessary to remove the temporary protection just installed to find the leak. Some leaks occur just under certain circumstances and cannot be found except under those same circumstances. Some leaks can be found by damming and flooding a portion of the roof.

Even while water is coming in, it may be difficult to find a roof leak. The source of the leak might not be exactly over the drip. Water can migrate between roofing layers or along the supporting structure before dripping. While the migration may not be as far in a low-slope roof as in a steep roof, the distances can still be several feet.

After locating the source of the immediate problem, examine the entire roof and flashings for similar damage. If the damage results in any way from the life of the roofing, or if there are many leaks or there have been many earlier leaks, perhaps the leak is only the first sign of a more general problem. When the failing roofing is at the end of its expected life span, replacing or re-covering the roofing may be in order.

When damage is apparent, the roof deck and structural supports should also be examined for damage.

Before new fasteners are driven into an existing deck, core samples should be taken to verify the depth and type of fastener needed, and the deck should be tested for pullout. At least ten tests should be made, most of them concentrated in perimeter and corner areas and in areas where leaks have occurred or wet insulation was found. More tests should be made if the first tests show low pullout capacity. The inability of a deck to hold fasteners is a sign that the deck has been damaged and may require repair or replacement.

Repair, Replacement, and Re-Covering Work—General Requirements

Compatibility. Membrane, adhesives, bitumen, solvents, fasteners, and other materials used to repair existing roofing or install new roofing must be compatible with the existing materials in every respect. Asphalt-based roof mastic should not be used to repair EPDM, for example, because EPDM reacts adversely to petroleum-based products. PVC is also sensitive to coal-tar pitch and asphalt. Neoprene, on the other hand, is compatible with asphalt. The manufacturer or contractor should take samples and have laboratory tests made to verify compatibility. Unless the materials are known absolutely, field examinations or tests will seldom be accurate enough to be considered reliable. Coal-tar can sometimes be tentatively distinguished from asphalt by shaking jars containing bitumen specimens and mineral spirits. A coal-tar sample will turn the liquid yellow, while asphalt will turn it black. All field tests should be laboratory-verified.

Sometimes, incompatible materials can be used in proximity if they are separated by a barrier that is compatible with both. An aluminum foil barrier is sometimes used between PVC and asphalt, for example.

Unfortunately, standards are not available for many roofing materials, especially single-ply membranes, and not all available standards are universally accepted by roofing industry members. The lack of standards, while frustrating, is understandable. The number of single-ply membrane blends, composites, and formulations available is staggering. In 1988, for example, there were more than 150 polymer-modified bitumen types on the market. So determining compatibility is sometimes difficult. Manufacturers change their products periodically or switch to a different material entirely. Old and new materials with the same name may have different formulations. There is no central source of materials data. New information and new products appear frequently, sometimes making textbooks and handbooks out of date before they are available for use. As a result, designers must become much more knowledgeable about roofing than most of them are comfortable with or run the risk of selecting the wrong materials or wrongly designing the system, with potentially damaging and expensive results to all involved, including the designer.

Code and Insurance Compliance. It is essential that materials and systems used in patches and new roofing comply with current fire and insurance requirements, even when the existing materials do not comply. It is also essential that the final in-place roofing system comply with code and insurance requirements. Failure to ascertain and, if necessary, obtain materials and system approvals from authorities having jurisdiction may create liability, lead to rejection of the roof by local authorities, or result in denial of insurance. It is not adequate to just use new materials exactly like those in place without verifying compliance. The existing materials may no longer comply with the code. The system, after new components have been added, may not meet fire-resistance requirements.

Particular care should be taken to determine if materials left in place, such as bitumens adhered to the roof deck, will affect fire ratings in the new systems. Materials which do not comply must be removed, regardless of the difficulty of doing so.

Weather Conditions.   Because of the danger of trapping moisture in the roofing system, roofing work should not be done during inclement weather. Substrates should be permitted to dry completely before new roofing is installed.

A secure night seal (cutoff) should be installed when roofing is temporarily stopped to prevent water from entering the system.

Repair versus Replacement versus Re-Covering.   ASTM defines replacement as removing the existing roofing and providing new roofing in its place. Re-covering is placing a new roofing system over an existing roofing system without removing the existing system.

Refer to Chapter 5 for a discussion about installing new insulation over existing roofing, whether wet insulation should be removed, and whether wet insulation can be dried using roof vents. The conclusion stated there is that wet insulation cannot be dried out using vents or other means and should be removed completely.

Deciding whether to repair an existing roof or to re-cover or replace it requires weighing the answers to the following questions:

Has the roofing reached its expected life span?

Is the damage general in nature or limited in scope?

Is the supporting construction damaged and in need of repair?

Is the insulation wet or otherwise damaged?

When the answers to those questions suggest that repairs are not the appropriate response to a roofing problem, a decision must be made whether to re-cover or replace the existing roofing, which is a difficult decision under the best of conditions. The decision requires weighing the answers to the following questions:

- Is it possible to prepare the existing surface to yield a satisfactory (smooth and strong enough) surface to receive the new roofing and to support the loads associated with installing the new roofing?

  The answer is different for different roofing materials and different degrees of damage. The condition of an existing material should be good enough so that it does not interfere with proper installation of the new material. Existing insulation must be properly fastened in place and provide sufficient impact resistance for the loads to be applied and the traffic necessary to install the new roofing.

  A competent roofing contractor, roofing consultant, or architect and the manufacturer of the roofing membrane to be used should be consulted at the site before a decision is made to install new roofing over existing roofing or insulation. "Competent," in this instance, means having knowledge of, and experience with, roofing problems in general and specific conditions similar to those involved.

  If existing roofing damage has advanced too far, removal may be necessary. If leaks or other conditions have saturated the insulation, or damaged the decking or structural supports, their repair or replacement may dictate that roofing should be removed.

- What will it cost to prepare the existing roofing to receive the new roofing?

  When the cost of preparing nears the cost of removing, re-covering becomes questionable.

- Will leaving the existing roofing in place yield a better completed roof than would removing the existing roofing?

  Re-covering has inherent in it the danger of trapping moisture in the existing roofing system. The potential harm may outweigh the savings made by not removing the existing roofing.

- Are the existing roof deck and structure capable of safely supporting the existing roofing, and new roofing, and the other loads associated with installation?

  Roofing over existing materials is advisable only if the substructure is strong enough to support the additional weight, including that of workers installing the roofing. Even when a structure appears to be sufficiently strong, it is a good idea to have a structural engineer examine it to be sure. Some problems may not be apparent to an untrained observer. For instance, a roof might be quite adequate to support the roof dead load and the live loads generally used at the time the roof was designed. But codes and engineering practices change.

What was considered adequate 50 years ago, when the building was designed, may violate current law.

- Is the risk to the building and its contents of tearing off the existing roofing before applying a new weather barrier acceptable?

A sudden, unexpected rainstorm, which can happen anywhere, that occurs while the roof is uncovered might severely damage an unroofed building's contents (see Fig. 7-8). Some contents, such as those in a museum or art gallery, may be too valuable to afford taking even a small risk.

- Is there already more than one roofing system in place?

Some codes limit the number of systems that can be superimposed. Generally, not more than two roofing systems should be applied on a roof.

When the decision is made to remove the existing roofing membrane, it is necessary to decide whether to remove the existing insulation or to install the new roofing over it. Refer to Chapter 5 for further discussion.

When the roofing is replaced or re-covered, existing composition flashings

Figure 7-8 Extensive damage to contents is possible if rain occurs during built-up roofing tear-off such as this. (*Photo courtesy of The Garland Co., Inc.*)

should be removed and new flashings installed. It might be necessary or desirable to also remove existing metal flashings when roofing replacement or re-covering is elected. Refer to Chapter 8 for additional discussion.

When the existing roofing is not to be re-covered or replaced, whether existing damaged flashings should be removed and replaced with new flashings or repaired depends on how badly they are deteriorated, whether they have reached their expected life span, and how widespread the damage is. The decision may be dictated by whether there is damage beneath the flashings that requires flashing removal. Such conditions include wet insulation, defective or missing cant strips, and substrate or structure damage.

Draining an Existing Roof.   Roofs pond water either because they were poorly designed or because they were deliberately designed to hold water. The practice of deliberate ponding for thermal or site drainage purposes, once in vogue in certain parts of this country, are generally discredited today as more troublesome than advantageous. Some manufacturers' literature still contains specifications for water-retaining roofs, however, and some existing roofs were designed to hold water. The best advice anyone can give an owner with a roof on which water ponds, whether designed that way or not, is to drain the roof permanently before it fails prematurely, unless there is some legal reason for not doing so (see Fig. 7-9).

When a roof ponds because the drains are of a type designed to restrict flow, draining the roof may be as simple as changing the drains. Often though, an existing roof which ponds water will be difficult to drain properly. There may be too few drains. The drains may be located so that proper drainage cannot occur. There may be little or no roof slope. When leaks begin to occur, however, draining is essential. Automatically cycling mechanical drainage systems are available but are limited in that they are powered by electricity or solar energy, both of which sometimes fail during rainstorms when they are needed most.

Often, the only effective long-term solution to a roof that does not drain properly is to make permanent changes in slope, or drain locations, or both.

Repairs after Blow-Offs.   Blow-offs occur because the existing materials were not properly anchored. It thus becomes necessary to determine the full extent of the damage. The portion of roof that has lifted and not returned to its original portion will be apparent, of course (see Fig. 7-10). But roofing and insulation often lifts and falls back into its original position so exactly that it is not apparent by visual examination alone. Therefore, it is necessary to make cuts and cores beyond the known damage to ascertain the exact extent.

Materials that have been separated from the roof deck by wind uplift should be removed. Membranes should be discarded. Insulation that has

Figure 7-9 Inadvertent ponding. This condition should be corrected before the roofing fails. (*Photo courtesy of The Garland Co., Inc.*)

been damaged in any way, including crushed corners or edges, or displays fracture lines should be discarded. New materials should be installed and properly anchored in place. Pullout tests should be conducted to determine if the deck will hold the new materials, and if necessary, the deck should be repaired or replaced as well.

Metal flashings that have suffered minimal damage but have not been bent back or blown off can sometimes be nailed back in place, using additional nails.

Gravel blow-off can be cured by replacing the gravel with a mixture that has larger aggregate, by substituting concrete pavers for the gravel, and, sometimes, by increasing the parapet height.

Repairing Existing Flashings.   Refer to Chapter 8 for a discussion of metal flashing repair.

Flashings beyond repair should be removed and new flashings installed. It may also be necessary to remove existing flashings for other reasons. New flashings to replace removed flashings should be installed just as new flashings would be installed in a new roof. Additional care must be taken, of course, to ensure the compatibility of new with existing materials and

Figure 7-10 Blown back EPDM. (*Photo courtesy of The Garland Co., Inc.*)

to prevent aggregate surfacing from becoming embedded into the new flashing. Materials used in new flashing application should be of a type acceptable to the roofing membrane manufacturer.

Standing water is often a major contributor to flashing damage (see Fig. 7-11). The cause of standing water should be eliminated and the flashing and adjacent materials allowed to dry before repairs are attempted.

In order to make base flashing repairs, metal cap flashings should be bent up out of the way. After base flashing repair, the caps should be bent back down into position covering the base flashing and straightened. Metal flashings loosened or removed to make base flashing repairs should be properly reinstalled (see Chapter 8).

Installing New Roofing over Existing Materials.   Before new roofing is applied over existing substrates, severely damaged or wet materials should be removed and replaced with new materials. Damaged decking should be repaired or replaced. Loose decking should be fastened securely. Existing ballast should be completely removed. Blisters, ridges, and other surface projections should be cut down. Loose felts should be removed. Dirt, dust, debris, and other foreign substances should be removed from all surfaces. Proper fastening of existing materials to the deck should be verified. Existing

Figure 7-11   Standing water along a flashing line. (*Photo courtesy of The Garland Co., Inc.*)

materials that are not properly fastened should be fastened in the same manner as would be new materials of the same type. Mechanical fasteners, where the deck is steel or nailable, should be selected to accommodate the existing and new materials' thicknesses.

Existing flashings should be removed whenever new roofing will be installed, even when the existing roofing is not to be removed. New flashings should be installed in accordance with details appropriate for the conditions. New flashings should be used in every location where flashing occurred in the existing roof and in other locations where flashing is appropriate.

All existing surfaces to receive new bituminous materials should be primed with appropriate primers, except that coal-tar bitumen may not need priming.

Special precautions should be taken to ensure that existing aggregate surfacing does not get into the new roofing, where it can damage the membranes or create bitumen voids.

The National Roofing Contractors Association's 1986 *The NRCA Roofing and Waterproofing Manual* contains specifications and details for installing many types of roofing over various substrates, including some existing substrates. Roofing manufacturers also furnish specifications and details for installing roofing systems using their products. Most of the latter are intended for new construction, but most manufacturers will provide guidance for adapting their new-work specifications and details to work on existing roofs. In every case, whatever specification or details are used as a guide, it is necessary to make sure that the system, materials, and details used are appropriate for the condition.

Contract drawings should be prepared for every roofing re-covering and replacement project to ensure that all potential problem areas are considered and solved before the fieldwork begins.

Roofing over Existing Roofing.   Refer to Chapter 5 for a discussion of installing thermosetting insulating fill or sprayed-in-place polyurethane foam over roofing damaged to the extent that it cannot be re-covered using built-up or single-ply membrane systems or board insulation without removing the existing roofing. Also refer to Chapter 5 for a discussion of installing new board insulation over existing roofing, including the need to verify structural and roofing system designs and the possibility of changing the roof slope during the process.

When new insulation has been installed, the methods of applying new roofing membrane are the same as that which would be used for new roofing in a new building. Fasteners, of course, must be selected to accommodate the existing material thickness.

New roofing membranes should never be adhered directly to an existing membrane. Instead, a layer of ventilated base sheet or porous insulation should be placed between the two. On nailable decks the covering should be nailed in place. On nonnailable decks, the covering should be placed in a layer of compatible bitumen or adhesive. Where there is a high chance of trapping moisture in the existing roofing system, some sources recommend cutting or otherwise penetrating the existing membrane to permit vapor to escape into the venting base sheet or porous insulation layer. The venting base sheet or porous insulation layer should then be vented to the exterior using pressure-relief vents, edge venting, or both. Where the old membrane is not needed as a vapor retarder, some sources recommend drilling a 3/4-inch-diameter hole every 100 square feet to permit water that gets into the new roofing to drain, so that leaks can be detected.

Roofing over Existing Insulation.   Refer to Chapter 5 for a discussion of installing new insulation over existing insulation.

Sometimes, it is possible to remove the existing roofing while leaving the existing insulation in place. After proper preparation, installing new roofing is the same as installing new roofing over new insulation.

Roofing over Cleaned Decks.   Placing new roofing over an existing deck from which the original roofing has been completely removed is the same as installing the same material over a new deck.

Repairing Existing Built-Up Roofing

Repairable roofing damage can be classified as blisters, ruptures, punctures, splits, damaged joints (fishmouths, loosened edges, buckles, and open laps), alligatoring, membrane dry-out, bare spots, and membrane deterioration. When any one of such damage types is extensive, or when several damage types exist at once, it may be time to consider re-covering or replacing the roofing.

The following suggested repair methods are a compilation from several sources. Individual product manufacturers should be consulted before repairs are made. They may recommend different methods or different materials, such as steep asphalt or roofing mastic, in lieu of the bitumen used in the original roofing. In general, the recommendations of a reputable manufacturer should be followed, unless obviously incorrect.

The following recommendations do not include material weights or coverage rates, which differ with different products. Consult the manufacturer of the products selected for those requirements.

In roof repairs, care must be taken to ensure that aggregate surfacing material does not end up within the roofing system.

The following repair recommendations assume that the roof has aggregate surfacing. The major differences in the recommendations when the roofing is smooth-surfaced are:

Eliminate all reference to aggregate surfacing.

Use impregnated glass fiber mesh felts, regardless of the felts used in the original roofing.

Clean off and coat the entire roof, instead of just the repaired area, with asphalt emulsion. Use fibered or nonfibered aluminum if originally used on the roofing being repaired.

Blisters.   Blisters are air bubbles beneath the membrane or between plies. They may appear as visible raised bubbles or be apparent only as spongy areas. Blisters result from trapped moisture or incomplete bitumen coverage in the initial application. Blisters can be a sign that the underlying insulation is wet. Refer to Chapter 5 for a discussion of methods that can be used to detect wet insulation and measures that should be taken when it is found.

Blisters should be left alone unless they have lost their aggregate covering, exposing the membrane; are in danger of being broken because of heavy foot traffic; or have ruptured, allowing water to enter. Blisters that have lost their aggregate cover should be treated as is recommended below for treatment of bare spots.

When small blisters have ruptured or must be repaired for other reasons, the following technique should be used:

1. Remove all aggregate from the area within 2-1/2 feet of the blister in all directions and sweep that area clean of dust, dirt, and other foreign substances.
2. Make two cuts through the blister at right angles to each other and turn back the edges. Remove water and foreign materials and allow the area beneath the felts to dry.
3. Prime the area beneath the cut membranes, except that priming may not be necessary for coal-tar bitumen. Trowel-apply roofing cement over the entire area and press the segments back into place with a roller, expelling all air.
4. Coat the entire cleared area with bitumen of the same type used in the original roofing and apply a roofing ply sheet to lap the patched area by at least 12 inches in all directions. A single-layer patch of modified bitumen may be used in lieu of the roofing ply sheet patch.
5. Apply a flood coat of bitumen and reapply the removed aggregate or apply new matching aggregate to completely cover the area where aggregate was removed, leaving no voids or bare spots. Existing aggregate, if reused, should be sifted or washed to remove excess silt and dirt.

Large blisters that have ruptured should be treated as is suggested below for ruptured membranes.

Ruptured Membranes.   Membranes rupture from several different causes, including broken blisters. Ruptured membrane should be repaired as follows:

1. Remove aggregate covering at least 2-1/2 feet in all directions beyond the area to be cut out. Sweep the area clean of dirt, dust, and other foreign materials.
2. Cut and remove damaged membranes down to sound material and beyond the damaged area to a dry membrane at least 2-1/2 feet beyond the damaged area in all directions. Leave no loose felts. Remove water and foreign materials and allow the area from which the felts were removed to dry.
3. Prime the area beneath the removed membrane, except that priming may not be necessary for coal-tar bitumen.
4. Install in the cut-out area one more than the number of felt plies removed, using the same ply type and bitumen that was used in the original roofing. Extend the bottom ply at least 6 inches beyond the

cut-out area and each succeeding ply 3 inches beyond the preceding ply. Press plies into place using a roller, expelling all air. A single-layer modified bitumen patch may be used in lieu of the multilayer patch.

5. Apply a flood coat of bitumen and reapply the removed aggregate or apply new matching aggregate to completely cover the area where aggregate was removed, leaving no voids or bare spots. Existing aggregate, if reused, should be sifted or washed to remove excess silt and dirt.

Punctures.   Punctures may result from fasteners backing out of decks or dropped objects. Small punctures can be temporarily patched with a layer of felt embedded and topped with flashing cement. Permanent puncture repair should be made as follows:

1. Remove aggregate covering at least 2-1/2 feet beyond the puncture in all directions. Sweep the area clean of dirt, dust, and other foreign materials.
2. Prime the area from which the aggregate has been removed, except that priming may not be necessary for coal-tar bitumen.
3. Install two plies of the same felt used in the original roofing over the cleared area using the same bitumen that was used in the original roofing. Press the plies into place with a roller, expelling all air. A single-layer modified bitumen patch may be used in lieu of the two-ply patch.
4. Apply a flood coat of bitumen and reapply the removed aggregate or apply new matching aggregate to completely cover the area where aggregate was removed, leaving no voids or bare spots. Existing aggregate, if reused, should be sifted or washed to remove excess silt and dirt.

Splits.   Splits are caused by expansion or contraction in the membrane, the drying out of the felts, felt slippage, curled or cupped insulation (see Chapter 5), and low tensile strength in the membrane. Before splits are repaired, the underlying cause should be found and eliminated. If not, repaired splits may open again, or the roofing will split near the original split. Repair splits as follows:

1. Remove aggregate covering at least 2-1/2 feet beyond the split in all directions. Sweep the area clean of dirt, dust, and other foreign materials.
2. Cut and remove damaged and loose membranes. Extend the split 12

inches at each end by cutting the membrane. Make a 2-inch-long crosscut at each end to prevent the cut from extending spontaneously. Remove water and foreign materials and allow the area to dry.

3. Prime the area beneath the removed membrane, except that priming may not be necessary for coal-tar bitumen.
4. Over the split, cut, and removed plies, apply a slip sheet of base sheet or other suitable materials. Extend the slip sheet about 4 inches beyond the split, cut, and removed materials.
5. Install the same number of felt plies as was used in the original roofing. Use the same type of ply and bitumen that was used in the original roofing. Extend the first ply at least 6 inches beyond the slip sheet. Extend each succeeding ply 3 inches beyond the preceding ply. Press plies into place with a roller, expelling all air. A single-layer modified bitumen patch may be used in lieu of the multiply patch.
6. Apply a flood coat of bitumen and reapply the removed aggregate or apply new matching aggregate to completely cover the area where aggregate was removed, leaving no voids or bare spots. Existing aggregate, if reused, should be sifted or washed to remove excess silt and dirt.

Fishmouths, Loosened Edges, Buckles, and Open Laps.   Such conditions should be repaired as follows:

1. Remove aggregate covering at least 12 inches beyond the defect in all directions. Sweep the area clean of dirt, dust, and other foreign materials.
2. Cut the damaged area. Remove water and foreign materials and allow the area to dry. Press the cut felts down until flat in a bed of the same bitumen used in the original roofing.
3. Prime the area to be patched, except that priming may not be necessary for coal-tar bitumen.
4. Over the damaged area, apply two plies of felt. Use the same type of ply and bitumen that was used in the original roofing. Extend the first ply at least 6 inches beyond the damaged area and the second ply at least 6 inches beyond the first. Press plies into place using a roller, expelling all air. A single-layer modified bitumen patch may be used in lieu of the two-part patch.
5. Apply a flood coat of bitumen and reapply the removed aggregate or apply new matching aggregate to completely cover the area where aggregate was removed, leaving no voids or bare spots. Existing ag-

gregate, if reused, should be sifted or washed to remove excess silt and dirt.

Alligatoring and Membrane Dry-Out.   Alligatoring is a topping bitumen (flood coat) defect, characterized by multiple cracks, in no discernible pattern, ranging in width from hairline to pencil size. The effect is caused by the membrane drying out from exposure to sunlight. In aggregate-surfaced roofing, alligatoring occurs most often in bare spots.

Recoating does not cure alligatoring. The new coating simply proceeds to alligator itself. The thicker the coating, the more tendency there is to alligator.

Many roofing experts and manufacturers believe that in roofs that are essentially sound, alligatoring can be repaired with resaturants, which seal over cracks and imperfections and do not themselves alligator. Resaturants can also give extended life to dried-out felts.

As is the case with many subjects, not all manufacturers or roofing experts agree about the use of resaturants. Opponents say that resaturants can damage the roofing and after a few years force removal of the roofing instead of just re-covering, as may have been possible if resaturants had not been used.

A resaturant probably should not be used on a roof that displays many splits, blisters, or extensive membrane deterioration, or where the insulation or substrates are wet or otherwise damaged.

The manufacturer's instructions about which resaturant product to use and application rates and methods should be followed.

Bare Spots.   Bare spots are areas where aggregate ballast and surface bitumen are missing, exposing the membrane (see Fig. 7-12). Bare spots are caused by improper application of bitumen or migration of the bitumen after application. Bare spots should be repaired as follows:

1. Remove remaining loose aggregate from the bare area. Apply a thin coat of asphalt primer to the existing surfaces to be covered. Primer may not be needed where coal-tar bitumen is used.
2. Coat bare areas with hot bitumen.
3. Apply the removed aggregate, or apply new matching aggregate to completely cover the area where aggregate was removed, leaving no voids or bare spots. Existing aggregate, if reused, should be sifted or washed to remove excess silt and dirt.

Membrane Deterioration.   When large areas of membranes become deteriorated, especially if the expected life of the roofing has been reached, consideration should be given to re-covering or replacing the roofing. The

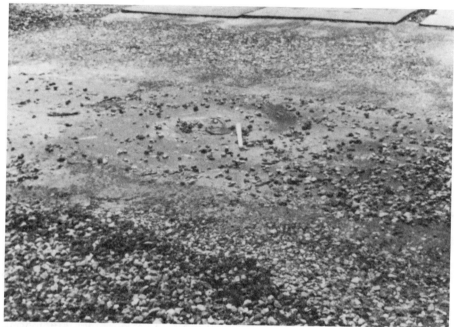

Figure 7-12   A bare spot caused by inadequate gravel adhesion. (*Photo courtesy of The Garland Co., Inc.*)

general condition of the roofing should also be considered. When deteriorated membrane is limited to small areas, the roofing is generally in good condition, the covered insulation is not damaged or wet, and the roof has not reached its expected life span, repairs can be made as follows:

1. Remove aggregate covering at least 2-1/2 feet beyond the deteriorated membrane. Sweep the area clean of dirt, dust, and other foreign materials.
2. Cut and remove damaged and loose felts. Remove water and foreign materials and allow the area to dry.
3. Prime the area beneath the removed membrane, except that priming may not be necessary for coal-tar bitumen.
4. Install the same number of felt plies as was removed, plus two. Use the same type of ply and bitumen that was used in the original roofing. Extend the first ply at least 6 inches beyond the edge of the removed plies. Extend each succeeding ply 3 inches beyond the preceding ply. Press plies into place with a roller, expelling all air. A single-layer modified bitumen patch may be used in lieu of the multiply patch.

5. Apply a flood coat of bitumen and reapply the removed aggregate, or apply new matching aggregate to completely cover the area where aggregate was removed, leaving no voids or bare spots. Existing aggregate, if reused, should be sifted or washed to remove excess silt and dirt.

Deteriorated roofing life may also be extended using a cold-applied polyester-reinforced coating system. Polyester membrane reinforcement can be coated with several elastomeric coatings.

Another method of extending roofing life is to cover it with an acrylic coating. Acrylic coatings form a fully adhered membrane directly over the existing roof.

Repairing Existing Built-Up Roofing Flashings

Deteriorated metal base flashings should be removed and replaced with bituminous flashings.

Repairable bituminous flashing damage can be classified as deteriorated surfaces and surface coatings; holes; open vertical laps and delaminated flashings; lack of bond between flashing and roofing; and separation from vertical surfaces. Figure 7-13 is an example of a repairable flashing condition.

The following suggested repair methods are a compilation from several sources. Individual product manufacturers should be consulted before repairs are made. They may recommend slightly different methods or different products. In general, the recommendation of a reputable manufacturer should be followed, unless obviously incorrect.

The repair recommendations that follow do not include material weights or coverage rates, which differ with different products. Consult with the manufacturer of the products selected for those requirements.

Before base flashings are repaired, aggregate surfacing should be stripped back from the area of the flashing at least 6 inches beyond the stripping onto the roofing. The flashing and the area from which the aggregate has been removed should be cleaned of loose particles, dust, dirt, and debris. After flashing repairs have been completed, a flood coat of bitumen should be applied and the removed aggregate reapplied, or new matching aggregate should be applied to completely cover the area where aggregate was removed, leaving no voids or bare spots. Existing aggregate, if reused, should be sifted or washed to remove excess silt and dirt.

Before asphalt bitumen is applied on existing materials, those materials should be primed. Coal-tar bitumen may not require priming.

Flashing patches should be coated with a layer of flashing cement. If

Figure 7-13   Failure along an expansion joint. (*Photo courtesy of The Garland Co., Inc.*)

the original flashing was coated with fibered or nonfibered aluminum, the patch should also be so coated.

Defects in bituminous base flashings may be repaired as follows.

Deteriorated Flashing Surfaces and Surface Coatings.   When surface coatings, such as flashing cement or aluminum fibered or nonfibered material, covering bituminous flashings have eroded away, clean the flashing, remove remaining loose coating material, and recoat the flashing. Use the same material used in the original coating.

The granules on mineral-surfaced cap sheets used in flashings will eventually loosen and fall off, especially if water stands on the flashing. Missing granules will allow the felts to absorb water. Flashing from which granules are missing should be coated with fibered aluminum after loose granules, dust, and debris are removed.

Holes.   Holes are caused by foot traffic, objects dropped onto the flashing, or deterioration of flashing materials. Traffic and dropped-object damage is common at base flashings, especially where cants have been omitted or are defective. When the decision is made to install new cants where existing cants are defective or missing, the existing base flashing must be completely removed. Then, new base flashing should be applied in the same way it would be for an all-new roof.

Holes in otherwise sound flashings may be repaired as follows:

1. A layer of flashing cement should be troweled over the hole to at least 6 inches beyond the hole.
2. Two plies of asphalt-impregnated glass fiber felt should be applied over the hole using flashing cement. The first ply should extend 3 inches beyond the hole in all directions and the second ply should overlap the first by at least 3 inches. A single-layer modifed bitumen patch may be used in lieu of the two-layer patch.

When holes result from flashing membrane deterioration and the decision is made to repair rather than replace the flashing, patching should proceed as follows:

1. Flashing cement should be trowel-applied over the hole extending at least 6 inches beyond the hole on each side and from top to bottom of the flashing.
2. Two plies of asphalt-impregnated glass fiber felt should be applied over the hole using flashing cement. The first ply should extend 3 inches beyond each side of the hole and from top to bottom of the flashing. The second ply should overlap the first by at least 3 inches on the sides and extend from top to bottom of the flashing. A single-layer modified bitumen patch may be used in lieu of the two-layer patch.

Open Vertical Laps and Delaminated Flashings. Loose and delaminated layers of flashing should be removed. If the underlying ply is damaged, it should also be removed.

Where flashing otherwise in good condition has separated at vertical laps, apply flashing cement beneath the loose material and press it back into place until flat. Cover the damaged area with a layer of felt embedded in flashing cement.

Lack of Bond between Flashing and Roofing. Stripping felts become loose because of differential movement between the stripping and the metal flanges of stripped flashing and specialty units. Some sources recommend repairing this condition in a manner similar to that recommended in this chapter for repairing fishmouths, loosened edges, buckles, and open laps in built-up bituminous membrane roofing. Others consider that procedure cumbersome and ineffectual. They recommend repairing composition flashing as follows:

1. Existing stripping should be removed and the exposed flanges cleaned and securely fastened in place if they are loose.
2. Two plies of stripping felts should be applied over the flange and adjacent roofing using roofing cement compatible with the existing bitumen. The first ply should extend not less than 3 inches out onto the roofing. The second ply should lap the first by 3 inches. A single layer of modified bitumen sheet may be used in lieu of two-layer stripping.

Separation from Vertical Surface. Base flashings sometimes fall away from walls and curbs, usually because of improper fastening, but sometimes because the wall or curb was not primed before the bitumen was applied. Separated membrane may sag (slide downward) or buckle (fall completely away from the wall or curb). The surest way to ensure a good flashing job is to completely remove the separated membrane and provide new flashing. Sagging membrane can, however, sometimes be repaired as follows:

1. Press the sagging membrane back against the wall or curb and nail it in place to the mortar joints or wood nailer. Place nails along the upper edge not more than 8 inches on center. Into masonry joints, use 1-1/2-inch masonry roofing nails through discs, or use nails with integral caps. Into wood, use appropriate nails through discs or with integral caps.
2. Where the upper edge of the base flashing is not covered by the metal cap flashing, trowel on a coat of flashing cement and apply a ply of asphalt-coated reinforced glass fiber flashing felt from the bottom of the flashing to the line of the cap flashing receiver. Coat the new ply with flashing cement.

When it is possible to return buckled flashing to its original location without breaking the membrane, follow the above recommendations for sagging membrane. Membranes that break or crack when pulled back to their original locations should be removed and new flashing should be provided.

Diagonal Wrinkling. Diagonal wrinkling of flashing membranes is caused by differential movement between the roof and adjacent vertical surfaces, such as parapet walls, or failure to anchor the roofing to the deck. Wrinkling will eventually lead to tears and splits, which can be repaired. Unfortunately, there is no effective way to repair wrinkling, short of removing the base flashing, correcting the underlying cause of the wrinkling, and installing new base flashing.

Repairing Existing Single-Ply Roofing

Repairable roofing damage can be classified as blisters, wrinkles, ruptures, punctures, splits, damaged joints (fishmouths, loosened edges, buckles, and open laps), dislocated aggregate ballast, and membrane deterioration. When any one of such types of damage is extensive (see Fig. 7-14), or when several types of damage exist at once, it may be time to consider re-covering or replacing the roofing.

The membrane manufacturer should be consulted before repairs are made and its recommendations should be followed exactly. Usually, repairs are made using the same membrane materials as were used in the roofing. The method of applying the patches, however, may differ. In vulcanized elastomeric membranes, and in true thermoplastic membranes, patches are usually adhered using the same method as was used in initial joint sealing. That is possible because the vulcanized membranes were already cured when the initial joints were made, and true thermoplastic materials do not cure in the field. But while initial joints in new uncured (nonvulcanized) elastomeric membranes that are not thermoplastic may have been heat-

Figure 7-14 Water is trapped beneath this severely damaged membrane. Re-placement is necessary. (*Photo courtesy of The Garland Co., Inc.*)

welded or solvent-sealed, after those membranes have cured in the field, patches cannot be heat-welded but must be applied using contact cement.

Repairs in polymer-modified bitumen membrane are usually made using a single-layer application of the membrane used in the roofing. Patches are applied using the same material and method as was used in sealing the initial joints. The recommendations outlined in this chapter for repairing built-up roofing apply to repairing polymer-modified bitumen roofing, except, of course, all patches should be made using the single-ply polymer-modified bitumen membrane option.

Before single-ply roofing is repaired, aggregate surfacing, if any, should be stripped back from the area to be repaired and the area swept clean of dust, dirt, and other foreign substances. Care should be taken to ensure that aggregate does not get into the repairs where it can damage the membrane or prevent complete bond. After repairs have been completed, the removed aggregate should be reapplied, or new matching aggregate applied, to completely cover the area where aggregate was removed, leaving no voids or bare spots. Existing aggregate, if reused, should be sifted or washed to remove excess silt and dirt.

Blisters are air bubbles in joints or beneath the membrane. They may appear as visible raised bubbles or be apparent only as spongy areas. Blisters result from trapped moisture or incomplete bitumen or adhesive coverage in the initial application. Blisters in adhered membrane can be a sign that the underlying insulation is wet. Refer to Chapter 5 for a discussion of methods that can be used to detect wet insulation and measures that should be taken when it is found. Blisters should be left alone unless they are in danger of being broken from heavy foot traffic, or have ruptured, allowing water to enter.

Membranes rupture from several causes, including broken blisters.

Punctures may result from fasteners backing out of decks or dropped objects (see Fig. 7-15).

Splits are caused by expansion or contraction in the membrane, drying out of the membrane, curled or cupped insulation (see Chapter 5), and low tensile strength in the membrane. Before splits are repaired, the underlying cause should be found and eliminated. If not, repaired splits may open again, or the roofing will split near the original split.

Fishmouths, loosened edges, buckles, and open laps occur usually because of improper sealing during installation (see Fig. 7-16).

Alligatoring is a defect in topping bitumen (flood coat), such as might be used on a polymer-modified bitumen membrane roofing system. It is characterized by multiple cracks, in no discernible pattern, ranging in width from hairline to the size of a pencil. Refer to "Alligatoring and Membrane Dry-Out" under "Repairing Existing Built-Up Roofing" for additional discussion.

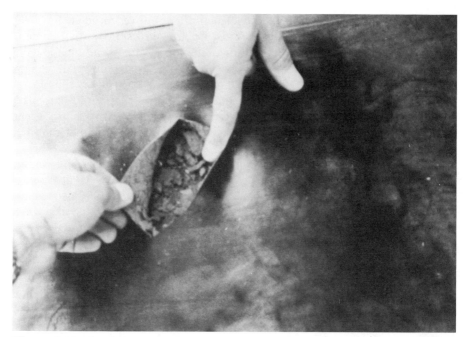

Figure 7-15 Punctured membrane. (*Photo courtesy of The Garland Co., Inc.*)

Dislocated aggregate should be swept back into its proper location. Consideration should be given to increasing the weight of aggregate, switching to concrete pavers, or increasing parapet height to prevent recurrence.

When membrane deterioration is widespread, especially if the expected life of the roofing has been reached, re-covering or replacing the roofing should be considered. The general condition of the roofing should also be considered. When deteriorated membrane is limited to small areas, the roofing is generally in good condition, the covered insulation is not damaged or wet, and the roof has not reached its expected life span, the deteriorated membrane should be removed and replaced with new membrane.

Sometimes, it may seem appropriate to reinstall undamaged portions of damaged loose-laid, partially adhered, or mechanically fastened roofing membranes, especially when the problem is caused by improper installation. For example, membrane with broken edge seals and joints caused by stretching during initial application might be reinstalled without stretching. Wrinkles, not caused by absorbed moisture, might be eliminated by raising the membrane, removing the cause of the wrinkles, and re-laying the membrane. Membranes wrinkled because they have absorbed moisture are damaged and should probably be discarded. But, properly reinstalling removed membrane may not be as easy as it appears. Joints, for example, are hard to

Figure 7-16 Open joint due to improper sealing during application. (*Photo courtesy of The Garland Co., Inc.*)

make. Repaired and reinstalled roofing may have a shorter life and generate more problems than new material. Money saved in materials may be spent several times over if the reinstalled roofing fails prematurely. Professional help and the manufacturer's advice is definitely needed before making the decision to reinstall removed single-ply roofing materials.

Repairing Existing Single-Ply Roofing Flashings

Deteriorated metal base flashings should be removed and replaced with elastomeric flashings compatible with the roofing membrane. In some membrane types, such as PVC, it may be possible to replace damaged metal flashings with new metal flashings which have been coated with the same materials as the membrane.

Repairable flashing damage can be classified as deteriorated surface coatings, holes, open vertical laps, separation from vertical surfaces, lack of bond between flashing and roofing, and, sometimes, wrinkles.

Product manufacturers should be consulted before repairs are made. In general, the recommendation of a reputable manufacturer should be followed, unless obviously incorrect. The same principles discussed above for patching single-ply roofing apply to membrane flashings.

Before base flashings are repaired, aggregate surfacing, if any, should be stripped back from the area of the flashing and out onto the roofing. The flashing and the area from which the aggregate has been removed should be cleaned of loose particles, dust, dirt, and debris. After flashing repairs have been completed, the removed aggregate should be reapplied or new matching aggregate applied to completely cover the area where aggregate was removed, leaving no voids or bare spots. Existing aggregate, if reused, should be sifted or washed to remove excess silt and dirt.

When surface coatings, such as flashing cement or aluminum fibered material, have eroded away, clean the flashing, remove remaining loose coating material, and recoat the flashing. Use the same material used in the original coating.

Holes are caused by foot traffic, objects dropped onto the flashing, or deterioration of flashing materials. Traffic and dropped-object damage is common at base flashings, especially where cants have been omitted or are defective. Where the decision is made to install new cants where existing cants are defective or missing, the existing base flashing must be completely

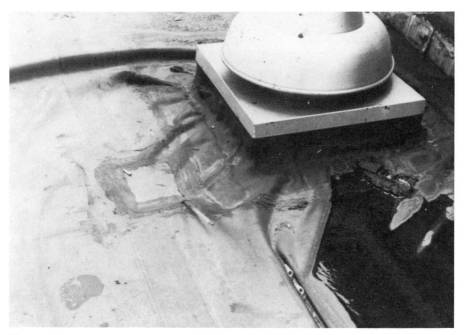

Figure 7-17 Wrinkled elastomeric flashing and delaminated patch. (*Photo courtesy of The Garland Co., Inc.*)

removed. Then new base flashing should be applied in the same way it would be for an all new roof.

When damage results from flashing membrane deterioration, the deteriorated membrane should be removed and new flashing installed.

Loose or severely wrinkled flashing should be removed (see Figs. 7-17 and 7-18). Uncorrected wrinkles will lead to tears and splits. Flashing that is damaged or nearing the end of its useful life should be discarded and new flashing installed. Flashing in good condition may be reinstalled using methods recommended by the manufacturer.

Base flashing sometimes falls away from walls and curbs, usually because of improper fastening, but sometimes because the wall or curb was not primed before the flashing was applied. Separated membrane may sag (slide downward) or buckle (fall completely away from the wall or curb). The surest way to ensure a good flashing job is to completely remove the separated membrane and provide new flashing. Separated membrane can, however, sometimes be reinstalled, following the manufacturer's instructions.

Stripping membrane sometimes becomes loose because of differential

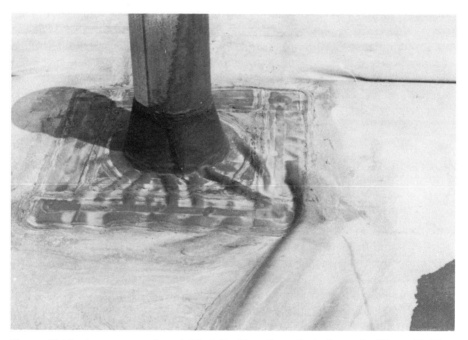

Figure 7-18 Less severely wrinkled flashing than that shown in Figure 7-17. This should be inspected frequently and repaired at the first sign of damage. (*Photo courtesy of The Garland Co., Inc.*)

movement between the stripping and the metal flanges of stripped flashing and specialty units or because of improper installation.

Diagonal wrinkling of flashing membranes is also caused by differential movement, this time between the roof and adjacent vertical surfaces, such as parapet walls, or failure to anchor the roofing to the deck. Wrinkling will eventually lead to tears and splits. Unfortunately, there is no effective way to repair diagonal wrinkling short of removing the flashing, correcting the underlying cause of the wrinkling, and installing new flashing. Sometimes, flashing that can be removed intact and undamaged can be reinstalled.

Installing New Single-Ply Roofing over Existing Roofing

Single-ply roofing can be installed directly over many types of existing roofing, including metal roofing, with proper preparation. Of course, the new roofing must be compatible with the existing roofing or properly separated from incompatible materials. Installation should be in accordance with the single-ply roofing manufacturer's recommendations, which should be verified with other roofing industry sources. Refer to "Where to Get More Information" and the Bibliography.

Where new insulation will not be used, a recovery board should be placed over properly prepared existing aggregate-surfaced, mineral-surfaced, and smooth-surfaced roofing before new single-ply roofing is applied. The recovery board may be a layer of insulation or another material recommended by the membrane manufacturer. The next step depends on the system selected and the method of application. A torch-applied modified bitumen might require a glass fiber mesh base sheet, for example.

Where to Get More Information

The joint Asphalt Roofing Manufacturers Association (ARMA) and National Roofing Contractors Association's (NRCA) 1988 publication *Quality Control Recommendations for Polymer Modified Bitumen Roofing*, while not an industry standard, contains recommendations for quality of workmanship, inspection procedures, substrate preparation, membrane installation, and flashing for new styrene butadiene styrene (SBS) and atactic polypropylene (APP) modified bitumen roofing. It covers heat-welded, hot-mopped, and adhesive, including partially adhered, applications. The publication does not address evaluating existing roofing but does cover installing new roofing over existing. Contact either ARMA or NRCA for copies.

The National Roofing Contractors Association's 1985 publication *Quality Control in the Application of Built-Up Roofing* contains similar requirements for built-up-roofing, but it has not been fully accepted by ARMA and has been criticized by some roofing products manufacturers, contractors, and roofing experts. One major disagreement concerns the 25 percent tolerance *Quality Control in the Application of Built-Up Roofing* permits for interply bitumen weight. Traditional tolerance has been 15 percent. Supporters say that the current equiviscous temperature (EVT) concept for controlling the bitumen's temperature, and thus its application viscosity, not only allows the tolerance to be lowered but forces the lowering because the bitumen is a liquid at its EVT and will be displaced more in one portion of the roof than in another because of felt roll weight. Thus, a test cut made at the location where a roll of roofing was first laid down would show less bitumen than a cut made further along in the roof. So, a roof could be rejected because of a condition that only occurs where rolls are first laid down, while the remainder of the roof might have even more bitumen than required. Opponents say that if a certain weight of bitumen is needed, it is needed throughout, and that the new standards were adopted to permit bituminous built-up roofing to compete with new roofing types.

In 1987, the Asphalt Roofing Manufacturers Association released its own *Built-Up Roofing Systems Design Guide,* which disagrees with the NRCA recommendations somewhat. NRCA, in turn, decided not to endorse the ARMA document. Probably the only solution for someone trying to reconcile the differing opinions in the roofing industry about built-up roofing is to obtain copies of the NRCA and ARMA recommendations, compare them, and make a decision based on which seems best for the problem at hand.

The joint national Roofing Contractors Association (NRCA) and Asphalt Roofing Manufacturers Association's (ARMA) 1988 *NRCA/ARMA Manual of Roof Maintenance and Repair* is an excellent guide to inspection, maintenance, and repair of low-slope roofs.

C. W. Griffin's 1982 book *Manual of Built-Up Roofing Systems* includes a comprehensive discussion of low-slope roof design and failure. Anyone with a roofing problem and every roof designer should obtain and read a copy of Griffin's excellent book.

Look for guidance about roofing fasteners and rated roofing systems in Factory Mutual System's (FM) *Approval Guide* and *Loss Prevention Data Sheets,* and in the Underwriters Laboratories' (UL) *Building Materials Directory—Class A, B, C: Fire and Wind Related Deck Assemblies* and *Fire Resistance Directory—Time/Temperature Constructions.*

The National Roofing Contractors Association's (NRCA) and 1986 *The NRCA Roofing and Waterproofing Manual* contains specifications and installation diagrams for many built-up and single-ply roofing installations, including applications over existing roofing.

The Single Ply Roofing Institute's 1987 publication *Single Ply Roofing*

Systems: Guidelines for Retrofitting Existing Roof Systems has advice on dealing with existing single-ply roofing.

For a detailed discussion of fastener failure, refer to Riaz Hasan's 1987 article "Eliminating Backout and Pullout of Roof Fasteners."

At the time of this writing (1988) The Single Ply Roofing Institute had published the valuable guide *Wind Design Guide for Ballasted Single Ply Roofing Systems* and was promising similar guides for mechanically attached and fully adhered systems. Every roofing designer should read the first publications, and if the other two are as valuable as the first, them also.

Stephen R. Hoover's 1987 article, "Single Ply Roofing: Exploring the Options," is an excellent account of the many different types of single-ply roofing materials and systems currently in use. It includes material characteristics, installation requirements, and description of some types of failure that can occur in single-ply roofing.

Roofing membrane and insulation manufacturers provide specifications and installation details for their products, which should be reviewed and studied. In addition, it is well to seek their advice for specific problems not addressed in their literature. Some produce separate documents recommending procedures and products for making repairs, which they do not include in their standard literature packages but will make available on specific request. It is not appropriate, however, to use manufacturers' data, or data from other industry sources, for that matter, without comparing that data with recommendations by industry sources and texts written by roofing experts and verifying that the data and other recommendations actually apply to the requirements of the project at hand. Details, of course, must be modified to suit the actual conditions at the project. The roofing industry is replete with conflicting advice and differences of opinion about many subjects, especially about the means, materials, and methods that ought to be used to repair damage. It is a good idea to double-check everything with multiple sources before coming to any conclusion.

8

Flashing, Roof Specialties, and Sheet Metal Roofing

Flashings discussed in this chapter include wall flashing, low-slope roof flashing, and steep roof flashing. Flashing materials discussed include sheet metal, composites made by bonding two or more materials together, and elastic flashings.

Roof specialties discussed in this chapter include low-slope roofing specialties and steep roofing specialties.

Sheet metal roofing includes architectural formed-in-place and preformed metal roofing. Both field- and factory-finished materials are included. Structural metal roofing is not included.

Steep roofing and associated bituminous flashings are discussed in Chapter 6. Low-slope roofing and associated flashing of the same materials as the roofing and flexible materials compatible with the roofing are discussed in Chapter 7.

All terms used here and not defined have their usual meanings. Refer to the National Roofing Contractors Association's 1986 *The NRCA Roofing and Waterproofing Manual* for an extensive glossary of roofing terms. C. W. Griffin's 1982 book *Manual of Built-Up Roof Systems* also contains a glossary of roofing terminology.

This chapter discusses the principles involved in producing some common types of flashing, roof specialties, and metal roofing conditions one might find in an existing building. It is not the intent of this chapter to discuss every condition that might be encountered, provide specific installation details, or imply that one detailing method, specific detail, or source document is superior to another. For additional guidance, refer to the sources listed in "Where to Get More Information" at the end of this chapter and in the Bibliography. Keep in mind, however, that this chapter and all the listed sources combined do not cover every conceivable condition. Often, it is necessary to understand and apply the principles involved to determine why a particular flashing, roof specialty, or metal roofing condition has failed, and to decide how to repair the damage.

Flashing, Roof Specialty, and Sheet Metal Roofing Materials and Their Uses

Flashing may be metal, composites made from metal and other materials, or elastic materials. Roof specialties are either metal or flexible materials with metal or flexible flanges. Sheet metal roofing is made from a variety of metals.

Metal Materials

Flashing, roof specialties, and metal roofing have been made from lead and copper for centuries. Lead and copper are still used today, of course, but in recent years, other metals have given designers and roofers many more choices. Unfortunately, the wealth of materials available sometimes leads to using the wrong one. Materials used today include the following:

Aluminum: A light, bluish silver-white metal that is ductile and malleable. Aluminum, which is earth's most abundant metal, is especially resistant to oxidation.

Aluminized steel: Steel sheet coated with an alloy of zinc and aluminum.

Zinc: A crystalline metal, which is bluish white in color. Commercial zinc is brittle at ordinary temperatures but ductile when heated. Zinc sheets for roofing and other uses is an alloy of zinc, copper, and titanium.

Galvanized steel: Steel which has been coated with zinc, either electrolytically or by the hot-dip process.

Stainless steel: An iron alloy which contains more than 12 percent chromium, and sometimes nickel, magnesium, or molybdenum. Most stainless steel does not rust and corrodes only under extremely severe conditions. Stainless steel is available in either hard temper or soft temper.

Steel: An iron alloy that contains carbon in amounts up to 1.7 percent. Each different steel (alloy) has a different amount of carbon. Steel is malleable under some conditions.

Iron: A heavy, ductile, and malleable metal that combines easily with oxygen to form the iron oxide coating called rust. Unrusted iron is silver-white in color.

Cast-iron: Cast-iron is an alloy of iron, carbon, and silicon. Cast-iron is hard, brittle, and not malleable.

Tin: A bluish white crystalline metal, tin is malleable and ductile. Tin has a low melting point. Tinplate is a sheet of steel or iron which has been coated with tin.

Terne: An alloy of lead and tin, usually in a ratio of four parts lead to one part tin. Terneplate, which is the "terne" used in roofing, is sheet iron or steel coated with terne.

Lead: A soft, heavy, malleable, ductile, plastic (but not elastic) metal, which is bluish in color. Some lead roofs have been in constant use for several hundred years.

Copper: A reddish metal that is ductile and malleable, copper is one ingredient in bronze, the other being tin. Copper is also the base for so-called "yellow metal." Copper, like lead, has been used for roofing for centuries. Sheet copper for roofing and flashing is available either plain or coated with lead. Copper sheet is either soft or "roofing temper" or hard "cornice temper." The National Slate Association booklet *Slate Roofs* recommends that only soft temper copper be used for flashings.

Flexible Sheets

Flexible sheets include elastic sheets made from polyvinyl chloride or modified polymers. Polyvinyl chloride deteriorates when exposed to sunlight and is recommended by its manufacturers for use in concealed applications only. Some modified polymers, however, are recommended by their manufacturers for use in exposed locations.

Flexible sheets also include glass fiber cloth, impregnated with rubberlike asphalt compounds, and composites of asphalt-saturated felt bonded between two layers of asphalt-impregnated cotton or glass fiber fabric.

Elastomeric sheets, usually plain or reinforced chlorinated polyethylene, are used as roof expansion joint covers. Usually the joint covering portion is bonded to a layer of foam insulation.

Composites

Composite flashing sheets are made from more than one kind of metal or from metal bonded to another product. Many composites are available. Some of them are:

A sheet of copper or aluminum with a layer of cotton or glass fiber fabric bonded to each side

A sheet of copper, aluminum, or other metal coated on each side with a layer of rubberized asphalt compound or an elastomeric film

A sheet of lead bonded to a sheet of copper, bonded to a sheet of glass fiber reinforced kraft paper or fabric

A sheet of copper with a sheet of glass fiber fabric, kraft paper, or a rubberized fabric on one or both sides

A sheet of copper bonded to a sheet of lead and the two faced on one or both sides with one or more layers of rubberized fabric, glass fiber fabric, or reinforced kraft paper

One advantage of many composite flashing sheets is that the layers of fabric or paper protect the metal from corrosion due to galvanic action.

Most composite sheets are not intended for use where exposed. Some of the parts of some composites will deteriorate quickly if exposed to sunlight.

Composites for use as expansion joint covers usually consist of a reinforced, insulated chlorinated polyethylene joint cover bonded on each edge to a sheet of metal, usually galvanized steel, stainless steel, or copper.

Sheet Metal Finishes and Textures

Sheet metals, depending on the material, use, and location, may be factory-finished, field-painted, or left unfinished.

Flashings and specialties associated with prefinished metal roofing are often the same material as the roofing and finished to match the roofing.

Most flashing not associated with prefinished metals is not factory-finished. Usually, but not always, flashings and roof specialties exposed to view from the ground or from inside the building are painted.

Flashings and roof specialties that are not visible from the ground or inside the building are often left unpainted.

Copings, gravel stops, and fascias are often factory-finished, even when not associated with prefinished metal roofing. When used with single-ply roofing, such items are often coated or clad with the same material used in the roofing membrane (see Chapter 7).

Some roofing, flashing, and roof specialty metal materials do not require finishing other than that done for aesthetic purposes. Copper, lead, aluminum, stainless steel, zinc, and terne-coated stainless steel are examples of materials that do not require painting for protection. Left unpainted, copper will usually turn blue-green. Depending on atmospheric pollutants, copper may continue to darken to almost black. Lead turns white over time.

Galvanized steel can be left unpainted but will eventually rust. Painting will make it last longer. Paint adheres more readily when galvanized steel is allowed to weather at least 2 years before it is painted.

Other metals should be painted to protect them from corrosion. Terne coating over plain steel, for example, contains pinholes that will permit water to reach and rust the steel unless painted. Terne-coated steel should be painted on both sides. The terne coating on terne-coated stainless steel, which weathers to a soft gray color, also has pinholes, but the stainless steel will not rust, so while terne-coated stainless steel can be painted, painting is not essential.

When exterior metals exposed to the weather are painted, the paint is usually an oil- or alkyd-based paint. Primers and undercoatings vary from paint to paint and metal to metal. Some metal items are factory-primed.

Factory finishes include porcelain enamel, fluoropolymer, acrylic enamel, siliconized polyester, and the same materials used in single-ply roofing membranes. In addition, aluminum may be clear or color-anodized. Most available factory finishes are warranted for periods that vary from manufacturer to manufacturer and finish to finish. For example, a typical warranty period for fluoropolymer is 20 years; for baked-on acrylic enamel, 10 years.

Sheet metal textures vary from smooth to stucco embossed. Metal panels may be smooth, ribbed, or corrugated. Flashings to be built into masonry are usually deformed to bond with the mortar.

Requirements Applicable to All Sheet Metals

While all sheet metal material and applications are unique in some way, they share some characteristics.

Material Thickness.   Sheet metal thicknesses vary for each use. The references given in "Where to Get More Information" at the end of this chapter include charts and lists suggesting the thickness of materials to use for flashing, roof accessories, and metal roofing. Some of them are repeated in this text. Where there is a conflict, the recommendations of the authority representing the material, or code requirements, if any, should prevail.

Fasteners and Nailers.   With few exceptions, fasteners in flashings, roof specialties, and metal roofing should be made from the same material as the metal being fastened. One exception is fasteners for cleats in lead roofing, which should be hard-copper wire nails. In every case, fasteners should be noncorroding. Where fasteners cannot be the same material as the metal being fastened, they should be of a metal that will not cause galvanic corrosion when in contact with the material being fastened or the supporting material.

Sheet materials more than 18 inches wide should not be nailed. Lead sheets should not be nailed regardless of size. Narrow sheet metals should be nailed only on one edge. Nails should generally be located not more than 3 inches apart and about 1/2 inch from the metal edge and should be staggered.

Sheets 18 inches wide and wider should be held in place by cleats, spaced 12 inches on center normally. Continuous cleats should be used to support sheets laid on surfaces which slope less than 3 inches per foot and on even steeper slopes when the metal manufacturer so recommends. Continuous cleats should also be used, regardless of slope, where the sheet edge is subject to high winds. Cleats may also be used on small metal pieces where necessary to hold the metal down. Cleats should be not less than 2 inches wide by 3 inches long and of the same material and thickness as the metal being held. Cleats should be nailed in place using at least two nails and lock-seamed into the sheet metal being held.

Aluminum should be fastened in place using screws. Other metals may be fastened in place using screws or twisted or threaded nails. Screws are often used in roof specialties; nails are used in flashings. Fasteners should extend not less than 1-1/2 inches into solid wood.

Nailers for fastening flashing, roof specialties, and metal roofing should be pressure preservative treated wood.

Edges.   Sheet metal edges not attached to or locked into adjacent metal sheets should be hemmed for strength.

Joints.   Sheet metals are joined by welding, soldering, lapping, lock-seaming, butting-and-back-flashing, or riveting, depending on the metal and use.

All metals can be welded, but some, such as aluminum, require special equipment and heat and humidity conditions which can usually be obtained only in the shop. Welding thin metals, other than lead, is difficult under any circumstances. Seldom are flashings, roof specialties, or metal roofing materials welded in the field. Joints in lead are, however, sometimes welded.

Usually, sheet metals are joined by soldering with a solder of tin and lead. The ratio of tin to lead varies depending on the soldered metal. A flux is used to aid in soldering. The flux material depends on the material to be soldered. On copper, the flux is rosin. Acid flux is used on stainless steel, zinc, and galvanized metal. Copper, lead, stainless steel, galvanized steel, terne-coated stainless steel, and zinc may all be soldered. Aluminum should not be soldered.

Most metals must be pretinned before soldering. Before copper, for example, can be soldered, the edges must be tinned 1-1/2 inches on both sides and the seams thoroughly sweated with solder.

Lead and lead-coated copper should be scraped or wire-brushed to produce bright metal before soldering.

Except in aluminum, base flashings should be soldered. Cap flashings may be lapped or seamed and need soldering only in areas subject to deep snow or high winds.

Aluminum assemblies are often joined by shop welding. Welding aluminum in the field is impracticable. Most field joints in aluminum are lapped, lock-seamed, or butted and flashed behind with a flashing plate. Gutters and downspouts are often lapped and riveted. All joints in aluminum should be filled with sealant.

Expansion and Contraction.   Expansion and contraction are particular problems in sheet metal work because of the large differential between its thermal movement and that of its supports. Expansion and contraction joints should be placed not more than 32 feet apart in aluminum and not more than 40 feet apart in other metals. Additional joints should be provided where the distance between the last joint and the end of the run of metal is more than half the recommended expansion and contraction joint spacing.

Extruded aluminum copings, gravel stops, and fascias should have expansion and contraction joints at about 12 feet on center.

Other expansion and contraction problems are caused by differential movement between flashings and their supports, between flashings and the membranes they flash, and between two different substrate materials or construction beneath the same flashing. Each of those conditions is discussed in this chapter.

Flashing Receivers.   The receiver shown in Figure 8-1 is a friction-lock reglet designed for casting into concrete. Face-mounted receivers, through-wall interlocking receivers, stucco-application receivers, and other types are also often used. The opening in a reglet for use in masonry walls is, of course, horizontal, to fit in the joints. The reglet in Figure 8-1 is shown sealed with lead wool and sealant. Lead wedges are sometimes used in reglets to lock the cap flashing in, but the design of some reglets negates the need for wedges. Reglets are made from various metals and plastics. The material used should be compatible with the flashing.

Cap flashings are sometimes inserted into a raggle (cut slot) left by the

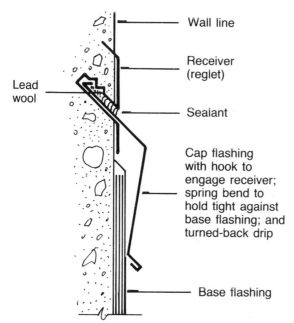

Figure 8-1 A typical cap flashing for concrete walls. Other types are available, but principles are the same regardless of details.

mason or cut into an existing joint and wedged in place using lead wedges or tension-forming shapes. The raggle is then filled with sealant. Raggles are more common where the flashing was added to an existing wall.

Cap flashings are sometimes lock-seamed into through-wall flashing or metal wall covering.

The National Slate Association's book *Slate Roofs* recommends that lead flashing never be calked directly into a reglet, but should be locked to a separate lead strip which is itself calked into the reglet.

Through-Wall Flashings

The purpose of through-wall flashing is to intercept water that may penetrate the wall's surface and lead that water back to the exterior. Therefore, all openings and penetrations through masonry and stone walls should be flashed, as should the tops of such walls and lines where structural supports or other members intercept the flow of water that penetrates the wall. Other types of walls also require flashing at similar locations unless the design of the wall automatically leads water that penetrates back to the exterior.

Except where joints are left open, joints in which through-wall flashing occurs must have weep holes through the mortar or sealant that closes them to allow trapped water to drain to the exterior.

Flashings should extend beyond opening jambs in every case and should be continuous at ledges and other water-catching lines.

The thermal expansion differentials between flashings and wall materials must be taken into account. Long runs of flashing must be broken into sections. In stone walls, it is sometimes necessary to break metal and composite flashing runs at spacings approximating the size of the stone pieces. Joints should be sealed to prevent leakage through the flashing. In stone walls, joints in the flashing should not fall in the same line as joints in the stone.

The following materials are used for through-wall flashings: stainless steel, plain copper, lead, lead-coated copper, zinc, galvanized steel, aluminum, elastic sheets, and combination products. Stainless steel through-wall flashings are often made from soft temper materials. Exposed stainless steel flashings subject to damage are usually made from hard temper material.

The Brick Institute of America's (BIA) 1985 *Technical Notes on Brick Construction* 7A says that copper and aluminum are likely to stain brick; that lead and aluminum will corrode in fresh mortar, and that galvanized steel is likely to corrode as well; that some flexible flashings are good, but others are not; and that asphalt-impregnated felt products should not be used in masonry walls. The recommended materials for unit masonry walls are copper, some plastics, stainless steel, and most combination flashings.

The Indiana Limestone Institute of America's *ILI Handbook* says that stainless steel, lead, zinc, copper, lead-coated copper, and aluminum can be used for flashing limestone, but cautions that copper and aluminum may stain stone. The same can be said for any other stone and for precast or cast-in-place concrete. Of course, the statements BIA makes about metal corrosion in fresh mortar apply to mortar used in concrete or stone. Stone and concrete installations should be designed such that water passing over flashings does not pass over the face of the stone or concrete.

Low-Slope Roof Flashings

Many flashings used in low-slope roofs are membranes of the same material as that used for the roofing or are similar or compatible membrane products supplied by the roofing manufacturer. Base flashings in bituminous built-up roofing are usually bituminous, for example. Most flashings in elastomeric roofing are supplied by the elastomeric roofing sheet manufacturer. Refer to Chapter 7 for a discussion of such flashings.

Flashings in metal roofing are usually the same materials as the roofing. Refer to "Metal Roofing" that follows for a further discussion of flashings used with metal roofing.

Cap Flashings. Cap flashings for low-slope roofs are usually either aluminum, stainless steel, copper, lead-coated copper, lead, terne-coated stainless steel, or galvanized steel. Shapes vary, but most have the basic characteristics shown in Figure 8-1.

Metal Base Flashing. Metal base flashings are not normally used in built-up or single-ply roofs because they are not flexible enough to mold to supports and because metals have expansion coefficients incompatible with most membrane roofing materials. Metal base flashings used with membrane roofing may be stainless steel, plain copper, lead, lead-coated copper, zinc, galvanized steel, or aluminum.

Roof Drains. Low-slope roof drains are usually flashed as follows: The roofing membrane is extended into the drain's flashing ring. A metal sheet about 30 inches square, usually of 4-pound lead or 16-ounce copper, is placed on the roofing membrane, extending into the drain's flashing ring at the center, and bedded in roofing cement. Then the metal pan is stripped in by the roofer using membrane strips.

The roofing cement used in the assembly must be compatible with the membranes and flashing. The same cement material will not necessarily be appropriate for different membrane materials. The manufacturer's recommendations should be followed.

The roof insulation is sometimes tapered down to the drain, and sometimes not.

Sometimes a metal gravel stop is installed around the perimeter of the drain flashing, forming a square about 36 inches on a side. Such a gravel stop is appropriate, however, only when the flashing material can be exposed to sunlight without damage or accelerated deterioration.

In some single-ply roofing systems, the single-ply membrane is simply extended into the drain's flashing ring and clamped.

Penetration Flashings. Probably the most talked about flashings, and those that give designers and building owners alike the most headaches, are flashings at penetrations through low-slope roofs. Appropriate details for some penetration flashings are generally agreed on. There is controversy, however, about other penetration details.

Flashings for single pipe vents through built-up roofs are handled in many ways, depending on the roofing material and system. Following are descriptions of some of the more common methods. Others may also occur.

Short pipes are often flashed with a sleeved flange, usually made of lead sheet. The flange portion should extend onto the roof at least 4 inches beyond the pipe in all directions and be flashed into the roof using membrane stripping. The sleeve should be soldered or welded to the flanges and should extend up the pipe for its full height and be turned down into the pipe at the top at least 2 inches.

In some types of single-ply membranes, the lead sheet is omitted and a flashing collar made from a compatible elastomeric material is dropped over the pipe and sealed to the roofing membrane and pipe using sealants. Sometimes a flashing drawband is used at the top to pull the flashing tight to the pipe. A sealant should always be used at the top of such flashing.

Some pipes and often conduit, shaped steel columns and pipes supporting equipment, and ducts may be flashed with two-piece formed metal housings. These housings come in many shapes, depending on the item being flashed, but share some characteristics. They have a flange which should extend at least 4 inches onto the roof all around the penetrating item and be flashed into the roofing using membrane strips. The sleeve may be shaped to fit closely around the penetrating item, or it may be rectangular or square with the penetrating item at the center. The sleeve should extend up the penetrating item pipe at least 8 inches above the roof. The sleeve is then capped with a metal hood attached to the penetrating item. The attachment may be by fastening to the item or, where the item has a regular shape, such as a pipe, by a drawband. On pipes, the hood is sometimes flanged out in an umbrella shape. In every case, the joint between the hood and the penetrating item should be sealed with an elastomeric sealant.

Some penetrations are flashed by passing them through a gooseneck or angled hood. The gooseneck or hood is flashed into the roofing using membrane strips.

Sometimes a curb is built around a penetrating item and flashed with base flashing. A cap is then fastened to the penetrating item by welding or is simply fastened to the item and sealed using sealant.

Curbs used to support equipment or devices are usually flashed using normal base flashing, which is then capped with metal counter flashing.

While all penetrations cause problems, the worst are irregular shaped or angled penetrations, such as guy wires and sloping rooftop-mounted sign supports. The National Roofing Contractors Association's 1986 *The NRCA Roofing and Waterproofing Manual* and most other sources recommend that pitch pans not be used. Unfortunately, it is not always possible to heed that advice. In addition, many existing buildings already have pitch pans. A good pitch pan, if there is such a thing, consists of a two-piece soldered metal pocket, set over the roofing membranes, flashed into the roofing by the roofer, and filled with a layer of some sort of mortar or a mixture of roofing cement and sand, and a layer of a self-leveling topping material. The recommended topping material varies from polyurethane sealant to coal-tar pitch. Certainly, both fill materials must be compatible with each other and with the roofing and flashing membranes.

Small penetrations also pose flashing problems. They are often flashed solely by turning the roofing membranes up on the penetrating item. Some small items are not flashed at all, except by a heavy coating of roofing cement. Both methods, unfortunately, will eventually fail. Small penetrations should be flashed, just the same as large projections, with flashed curbs.

Steep Roof Flashings

Steep roof flashings discussed here include those associated with the steep roofing types discussed in Chapter 6.

Many steep roof flashing applications are similar. There is little difference, for example, between valley flashing for wood shingles and shakes and valley flashing for mineral-surface cement tile. The suggestions that follow apply to all steep roofing, including metal roofing, except that metal roofing flashing is usually made from the same materials as the roofing. Refer to "Metal Roofing" later in this chapter for a further discussion of flashings used with metal roofing.

Steep roof flashings are formed from stainless steel, plain copper, lead, lead-coated copper, zinc, galvanized steel, or aluminum. Concealed stainless steel flashings are made from soft temper materials. Exposed stainless steel subject to damage is most likely hard temper material. Flashings should be compatible with the materials flashed and with the substrates.

Drip Edges.   Roll roofing and composition shingle roofs should have roof edge flashing with a drip at the eaves and along the rakes (see Fig. 8-2). Along the rakes, the drip edge flashing should be applied over the underlayment (see Chapter 6). Beneath roll roofing, the drip edge at the eaves should be placed over the underlayment. At the eaves beneath composition shingles, the drip edge should be installed directly on the deck, beneath the underlayment.

Base Flashings.   Base flashings along slopes in shingle-type roofing may be either pieced (see Fig. 8-3) or continuous (see Fig. 8-4). Pieced base flashing is woven with the roofing as the roof covering is placed. Each piece should be extended at least 2 inches beneath the next course of roof covering and should extend to within 1/2 inch of the lower edge (butt) of the covering piece of roof covering. Base flashing pieces should overlap in the direction of water flow by at least 3 inches.

Continuous base flashing is formed with a hook along the roof edge and fastened in place beneath the roof covering using cleats.

Both kinds of base flashing should turn out onto the roof and up the wall not less than 4 inches. (Some sources say that base flashing should turn up the wall at least 8 inches.) Where adjacent wall finishes are the

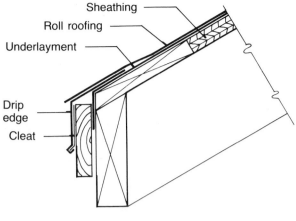

Sheathing

Roll roofing

Underlayment

Drip edge

Cleat

Roll Roofing Eave

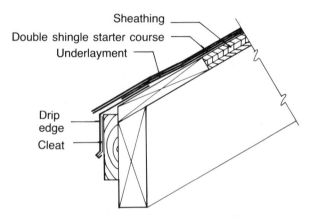

Sheathing

Double shingle starter course

Underlayment

Drip edge

Cleat

Composition Shingle Eave

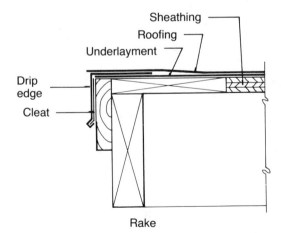

Sheathing

Roofing

Underlayment

Drip edge

Cleat

Rake

Figure 8-2 Drip edge details for roll roofing and shingles.

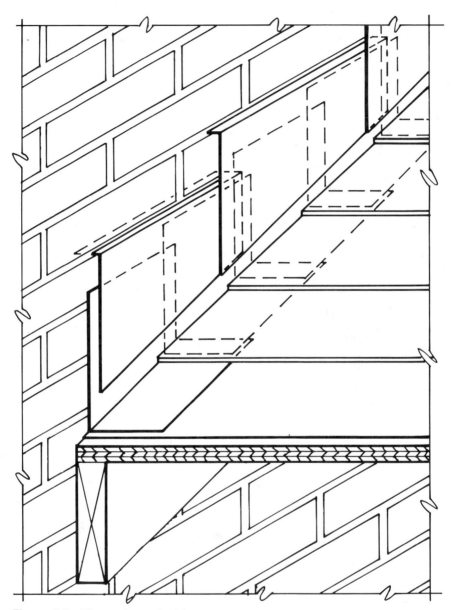

Figure 8-3 Woven base flashing.

cap flashing, base flashing should extend up behind those finishes not less than 6 inches.

Base flashings at walls where the roofing slopes away from the wall (see Fig. 8-5) should be placed as an apron with the vertical leg extending

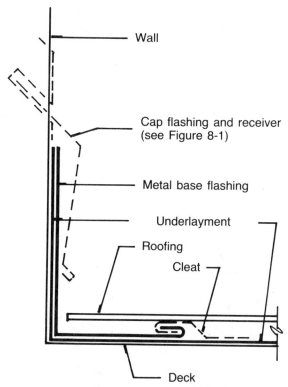

Figure 8-4 Continuous base flashing.

not less than 5 inches up the wall and the roof leg extending not less than 4 inches out onto the roof above the roof covering.

Cap Flashings. Some sort of cap flashing is necessary for every piece of base flashing, including apron flashing. Sometimes, the wall finish material acts as cap flashing, eliminating the need for using a separate piece of sheet metal for that purpose.

Figure 8-1 shows a typical type of cap flashing, but many cap flashings at steep roofing are quite different from that example. Sometimes, cap flashings will cover drip edges or extend beneath window sills, for example.

Along walls where the roof slopes, cap flashing is cut into sections (see Fig. 8-3). The length of the pieces will depend on the roof slope. No step should be more than three bricks high. Cap flashings at steep roofs should lap base flashings at least 4 inches.

Valley Flashing. Valleys in composition shingled roofs are sometimes

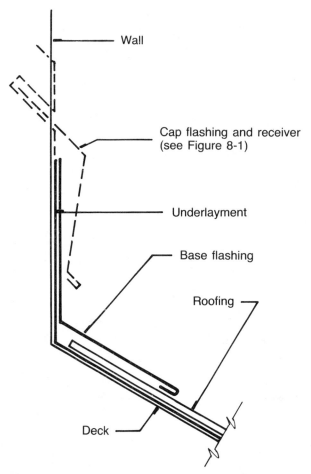

Figure 8-5 Top of sloped roof base (apron) flashing.

flashed with bituminous felts (see Chapter 6), and sometimes with metal. Valleys in other steep roofing are flashed with metal.

Open valleys flashed with metal are continuously flashed with long metal strips (see Fig. 8-6). Continuous valley flashing should be at least 18 inches wide at the top of the slope and taper toward the bottom at the rate necessary to keep metal always at least 11 inches beneath the roofing materials on both sides of the valley.

The center of continuous valley flashing should be crimped to prevent water rushing down the slope on one side from continuing up the opposite slope and under the roofing there.

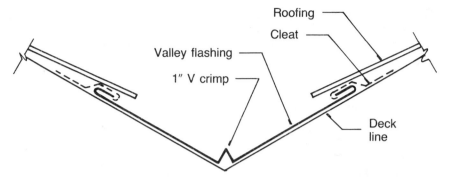

Figure 8-6 Valley flashing.

Continuous valley flashing strips should be lapped about 12 inches in the direction of water flow. The lap should be sealed with plastic roofing cement and not soldered. Flashing should be fastened in place using cleats along the edges at between 8 and 12 inches on center. The cleats should fall always beneath the roofing.

Most sources recommend that open valley flashing be applied over a felt underlayment. Some sources recommend that any valley flashing other than copper be painted on the underside to prevent damage due to condensation buildup beneath the metal.

Metal concealed valley flashings are short strips of metal woven into the roofing as the roofing is laid. Flashing sheets should extend at least 2 inches up the slope beyond the top of the roofing unit (shingle, for example) below. The bottom of the flashing sheet should stop 1/2 inch above the bottom (butt) of the covering roofing unit, so that the flashing is concealed. Concealed flashings should be 18 inches wide for slopes of 6 inches per foot and 24 inches wide for lower slopes.

Chimney and Other Major Roof Penetrations.   One method of flashing chimneys is to encase the chimney in what is called a stepped-pan through-wall flashing system, which is a one-piece flashing made by soldering metal strips into a shape that completely covers the profile of the chimney at a point about 8 inches above the roof, excluding, of course, an opening for the flue to pass through. The disadvantage of this system is that it has difficulty compensating for differential thermal and other movements between the chimney construction and the roof. Such a system is sometimes used nevertheless when the chimney materials are porous enough to possibly permit water to penetrate into the building through the chimney material itself.

Most chimneys and other major roof penetration such as skylights and dormers are flashed as follows:

The side walls along the slope are flashed using the kinds of vertical wall base and cap flashings described earlier (see Figs. 8-1, 8-3, and 8-4).

The lower side of major roof penetrations is flashed using apron flashing similar to that shown in Figure 8-5. The apron flashing should extend up the wall not less than 5 inches and out onto the roofing at least 4 inches.

The upper side of major roof penetrations should be flashed over a metal cricket (saddle). Joints in saddle flashing should be soldered. Saddle flashing should extend at least 4 inches up the wall and 4 inches out onto the roof beyond the cricket. Saddle flashing should not be nailed but should be fastened in place to the roof using cleats.

Minor Roof Penetration and Other Steep Roof Flashings.   Pipes through steep roofs are often flashed with a sleeved flange, usually made of lead. The flange portion should extend onto the roof at least 4 inches beyond the pipe in all directions. The sleeve should be soldered or welded to the flange and should extend up the pipe for its full height and be turned down into the pipe at the top at least 2 inches.

Very tall pipes through the roof may be flashed with a two-piece formed metal housing. The housing has a flange that should extend at least 4 inches onto the roof all around the pipe. The sleeve should extend up the pipe at least 8 inches above the roof. The top 3 inches of the sleeve should be capped by a metal hood attached to the pipe using a metal drawband. The joint between the hood and the pipe should be sealed with an elastomeric sealant.

In steep roofs, flashings are also necessary at changes in roof slope and over hips and ridges. Sheet metal water diverters are sometimes used over entrances in lieu of gutters.

Roofing Specialties and Accessories

The discussion here applies to devices the Construction Specifications Institute, Construction Specifications Canada's 1983 *Masterformat* calls roof specialties, including copings, expansion joint covers, gravel stops, fascias, gutters, and downspouts. The discussion here also applies to ridge vents and curbs, which *Masterformat* calls roof accessories. This book does not, however, include discussion of other items *Masterformat* calls roof accessories, such as gravity ventilators, skylights, heat and smoke vents, and roof hatches, except to say that they are items that require base flashings.

Roof specialties for use with metal roofing are discussed in the paragraphs of this chapter where metal roofing is addressed.

Roof Accessories.   Roof accessories for low-slope roofs often come with built-in counter flashing and roof flanges. Roof flanges usually extend out onto the roof about 4 inches for fastening in place through the roofing to nailers and for stripping in by the roofer. Base flashing is usually extended up the face of the accessory's curb and beneath the built-in counter flashing.

Roof accessories for use on steep roofs are fitted with their own flashing collars. The collars should extend out onto the roof deck at least 4 inches. Flanges are usually fastened in place using cleats. The accessory itself may have additional straps or other devices for fastening to the roof structure for additional support. Flanges for use with metal roofing should conform to the roofing profile.

Coping Covers.   Most metal coping covers are either formed sheet metal or extruded aluminum. Finishes are either mill finish or shop-primed for field painting or factory-finished. Copings are installed either with open joints or with joint covers. Coping covers are usually furnished in sections about 8 or 10 feet long.

Coping cover installation methods vary with the type of coping, but some characteristics are common. Most coping covers are installed over a layer of building felt. Outer edges are often locked in place using continuous edge strips (cleats). Inner (roof side) edges are sometimes fastened in place using screws and sometimes snapped over continuous edge strips. Some coping covers have a substrate support system; others are supported on wood nailers. Nailers should be pressure preservative treated.

Coping covers usually slope toward the roof so that water falling on the coping will not drain down the face of the building.

Expansion Joint Covers.   Expansion joint covers are either formed from metal or are prefabricated elastic products. Elastic expansion joint covers may have either elastic or metal flanges. Expansion joints may occur across the center of the roofing or where the roof abuts a vertical wall.

Metal expansion joint covers are either field- or factory-formed from one of the sheet metals generally used for flashing or roofing. Expansion joint covers may be factory-finished, but are often mill-finished or primed for field painting.

Metal expansion joint covers are usually installed on flashed curbs. The flashing joint cover acts as counter (cap) flashing.

Elastic expansion joints with bent metal flanges are often used on flashed curbs, which is the generally preferred method. The turned-down flanges act as counter (cap) flashing.

Elastic expansion joints with flat metal flanges or elastic flanges, although many roofing experts advise against such use, are sometimes installed on flat or tapered curbs or on nailers flush with the roofing. The flanges are flashed into the roofing. In some single-ply membrane roofing systems, the flanges are covered by a sheet of membrane which is sealed to the roofing and to the expansion joint cover.

Expansion joint covers for some single-ply membrane roofing systems are just a sheet of the membrane laid over flexible tubing. Sometimes, loose-laid membranes are installed directly over an expansion joint. These two methods are advisable only when the roofing manufacturer specifically recommends and agrees to warrant them and an independent professionally competent roofing expert concurs.

Gravel Stops and Fascias.   Gravel stops and fascias are usually prefabricated, either from sheet metal or from aluminum extrusions. Corners will also usually be prefabricated, and often mitered. Lengths of individual sections should be at least 8 feet. Roof flanges, where applicable, should extend at least 4 inches out onto the roof, and should be nailed in place.

Gravel stops should be nailed down over the installed roofing and then stripped into the roofing. Stripping-in methods and materials vary with the roofing type and material.

Most fascias should have hook strips at the bottom to positively hold the fascia in place. Some heavy extruded fascias do not require hook strips.

Gutters.   Gutters may be built into the roof on older buildings, occasionally even on newer buildings, but most gutters today are hung from fascias.

Built-in gutters are built using the same methods used to install any sheet metal construction. Of course, nails or screws should not be used in the water-carrying portion of the gutter. Built-in gutters may be built of any sheet metal normally used for flashing but are seldom built of aluminum. Aluminum cannot be soldered, which makes producing watertight joints even more difficult than usual. Seams in gutters must always be made to shed water in the direction of flow.

Hung gutters can also be formed of any normal flashing material, but most are aluminum, galvanized steel, stainless steel, or copper. Many shapes are used, but most are half round or ogee. Some hung gutters are supported from beneath by brackets, or internally using straps. Brackets and straps should be of the same material as the gutter, or of a compatible material. Bracket and strap size, thickness, and spacing depend on gutter size and material. Maximum support spacing is usually recommended as 30 inches on center, however.

Gutters can be set flat, of course, but a slight slope toward the drains is recommended to prevent water standing in the gutter.

In cold areas, hung gutters should be installed below the slope line so that snow and ice will slide clear.

Screens are advisable over downspouts to prevent leaves and other clutter from stopping up the downspout.

Expansion control, particularly important in gutter design, is accomplished in hung gutters by lapping one section of gutter over the next with a complete water barrier between the two sections. Another type of expansion joint amounts to completely stopping separate gutter sections on opposite sides of the joint and then covering the opening with a sheet metal cover to make the joint less conspicuous.

Expansion control in built-in gutters is a much more complicated undertaking, requiring expansion joints at the ends of the guttering and at intermediate points in the run of gutter. Detailing depends on the gutter design.

Scuppers. Scuppers may be constructed of any metal generally used for flashing. They consist of a sheet metal liner which is flashed into the roofing and which permits water to drain through a parapet or gravel stop without permitting water to flow into the building.

Scuppers sometimes drain into conductor heads, which are similar to short sections of gutter with both ends sealed. Conductor heads are usually connected to downspouts.

Downspouts. Downspouts draining built-in gutters are usually plumbing items beyond the scope of this book.

Downspouts for hung gutters are usually formed from the same metal as the gutter. Many shapes are used, but most are plain round, corrugated round, or corrugated rectangles. Size depends on downspout spacing, roof area drained, and geographical location.

Downspouts should be held at least one inch from the wall and supported by brackets, straps, or hangers of the same metal as the downspout, or of a compatible metal. Spacings of supports should be determined by the downspout size but should not be located more than 5 feet apart.

Downspouts may terminate with elbow-type fittings that drain onto splash blocks or into drainage lines. Terminations into drainage lines should be sealed with Portland cement mortar, which should slope away from the downspout.

Metal Roofing

Metal roofing can be either structural or architectural.

Structural metal roofing is just that, a structural roof which just happens to also form a weather barrier. Structural metal roofing is designed to span between structural roof supports, usually purlins. It does not require a roof deck. Corrugated roofing is an old form of structural metal roofing, the best known example of which is probably the World War II Quonset hut. Most modern-day structural metal roofs, however, are steel or aluminum panels joined by raised seams which provide stiffness and seal out the weather. Clips are used to attach the roofing to the structural supports.

Structural metal roofing can be used on low-slope roofs (slopes as low as 1/4 inch per foot). They can also be used to roof over an existing roof using a system of structural supports fastened through the existing roofing to the structure below.

Regardless of the materials or their use, structural roofs are beyond the scope of this book.

Architectural metal roofing has no structural function. It is installed over a complete structure, including a roof deck, and used solely as a weather barrier. When used on slopes lower than 3 inches per foot, architectural metal roofing must be specially handled to prevent water intrusion. Joints must be soldered or filled with sealant, for example. Architectural metal roofing may be either formed-in-place or preformed. The basic installation procedures are about the same for each. For the remainder of this chapter the term "metal roofing" and similar references refer to architectural metal roofing.

Materials. Metals for architectural metal roofing include lead, copper, lead-coated copper, terne-coated stainless steel, stainless steel, galvanized steel, zinc, and aluminum.

The National Slate Association booklet *Slate Roofs* states that the minimum weight for copper roofing and flashings is 16 ounces per square foot and that the best results are achieved using copper that weighs 20 ounces per foot. The Copper Development Association's 1980 publication *Sheet Copper Applications* recommends different weights of copper for different applications. For example, 16-ounce copper is recommended for standing seam roofing with 16-3/4-inch-wide pans and 20-ounce copper is suggested for standing seam roofing with 20-3/4-inch-wide pans.

Most sources recommend that lead roofing and associated flashings and roof specialties be formed from not lighter than 3-pound hard lead. Small pieces, however, might be made from 2-1/2-pound hard lead. Sometimes, batten seam roofing, where the battens are 18 inches on center or closer, is formed from 2-1/2-pound lead.

The American Iron and Steel Institute's 1972 publication *Stainless Steel: Suggested Practices for Roofing, Flashing, Copings, Fascias, Gravel Stops, and Drainage* recommends that 28 gage stainless steel be used for roofing with up to 18-inch-wide panels, 26 gage with panels up to 24 inches wide, and 24 gage for wider panels. Stainless steel for associated flashings and roof specialties varies from 22 gage for wide fascias to 28 gage for base, cap, and counter flashings.

The Sheet Metal and Air Conditioning Contractors National Association, Inc.'s (SMACNA) 1979 *Architectural Sheet Metal Manual* recommends that 26 gage galvanized steel be used for roofing panels up to 24 inches wide and that wider panels be formed from 24 gage material.

The SMACNA *Architectural Sheet Metal Manual* recommends 0.015-inch-thick terne plate be used for roof panels up to 24 inches wide, and that 0.178-inch-thick terne plate be used for wider panels.

Aluminum is seldom used to produce formed-in-place roofing. The most common preformed aluminum panel thickness is 0.032 inch. Thicker aluminum is used for copings, gravel stops, and other specialties.

Preformed zinc roofing is usually 0.027 inch thick but may be available in 0.020-inch thickness.

Preformed roofing usually includes panels with edges formed to suit the installation system to be used and appearance desired, and necessary accessories and fasteners. Preformed panels come in many shapes and sizes. Some have built-in ribs. Some have snap-on battens. Some are just pans with edges formed to permit field forming of standing seams. Some are roll-formed into corrugated shapes.

Unformed sheets for roofing may be unfinished or prefinished.

Sheathing.   Metal roofing should be installed over a nailable material, with sufficient nail holding capability to prevent nail pullout.

Underlayment.   Where the January mean temperature is 30 degrees Fahrenheit or less, an ice shield should be installed beneath metal roofing from the edge of the roof at the eaves to a line 24 inches inside the building wall. The ice shield should consist of two layers of 15-pound asphalt-impregnated, unperforated roofing felt installed using hot asphalt. As an alternative to the multiple layers suggested above, a single layer of an adhered membrane intended by its manufacturer to be used for that purpose may be used as the ice shield.

In addition, an underlayment should be provided beneath all metal roofing. The underlayment should consist of a layer of not less than 15-pound asphalt-impregnated, unperforated roofing felt. Edges should be lapped 2 inches. Ends should be lapped 6 inches. Sheets should be nailed in place using large-headed roofing nails.

In most cases, the felt should be completely covered by a layer of rosin sized building paper to prevent the roofing from sticking to the felt. Edges should be lapped 2 inches. Ends should be lapped 6 inches. Joints should not fall over joints in the felts. Sheets should be nailed in place using large-headed roofing nails of a noncorroding type compatible with the roofing.

The rosin sized building paper may have been omitted beneath metal roofing installed with flat seams since some designers do not demand that flat seam roofing be able to move.

Fasteners and Cleats.   Sheet metal roofing should never be fastened in place using exposed nails or screws. All fastening should be done through cleats, which are then locked into the roof covering material at the joints, concealing the fasteners from view and protecting them from exposure to the weather.

Fasteners should be of the same material as the roofing, or be of a compatible noncorroding material.

Generally, cleats should be of the same material as the roofing. Lead roofing, however, is usually installed using 16-ounce copper cleats, except that exposed cleats may be 3-pound lead. Usually, cleats are installed about 8 inches on center. On very steep lead roofs, horizontal cleats should be continuous.

Installation: General Requirements.   Each type of factory-formed roofing is installed in accordance with the particular manufacturer's recommendations. Installation methods may differ.

Most formed-in-place metal roofing is installed using either standing seams, flat seams, or batten seams.

All seaming methods can be used on slopes of 3 inches per foot and steeper. Flat seams can also be used on low slopes with sufficient pitch to permit proper drainage (no standing water). Joints in slopes less than 3 inches per foot, and in any slope subjected to high winds or other extreme exposure, should be welded (burned, in lead-trade terminology) when lead and soldered in other metal roofing materials. Where soldering is not practicable, such as in aluminum, joints should be filled with a nonhardening sealant. Some joints in other kinds of existing roofs may have been filled with sealant and not soldered. The safest course is to not use metal roofing on slopes lower than 3 inches per foot.

Standing seams should be not less than one inch high (see Fig. 8-7).

Flat-lock seams (Fig. 8-8) are made by interlocking adjoining sheets. Flat-lock seams are used as cross seams in standing seam and batten seam roofing and as all (except at expansion joints) seams in flat seam roofing. Most flat-lock seam roofing is laid with joints parallel and perpendicular to the eaves in a rectangular pattern. Some flat-lock seam roofing is laid in a diamond pattern where square roofing pans are laid with the joints at an angle (usually 45°) with the eaves.

In batten seam roofing, the sheets are joined over wood battens (see Fig. 8-9). The battens in batten seam roofing may run up and down the

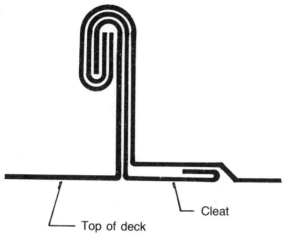

Figure 8-7   Standing seam.

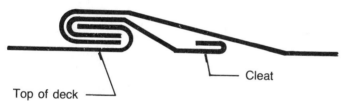

Figure 8-8   Flat-lock seam.

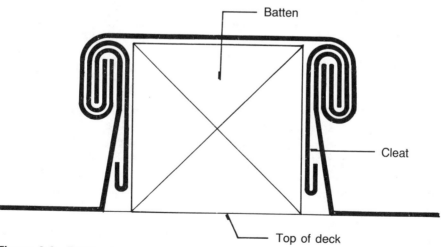

Figure 8-9   Batten seam.

roof slope from ridge to eave or form ornamental patterns, such as the vee-shaped layout called "chevron."

Other roofing patterns are also sometimes used. A "Bermuda" roof, for example, consists of a series of steps, resembling continuous shingles, which are formed in wood and covered with metal roofing. Bermuda roofing should not be used on slopes less than 2-1/2 inches per foot.

Expansion joints should be provided in metal roofing. Joint spacings vary depending on the roofing material and seaming method. In general, where ends are fastened, joints should be at 15 feet on center. Where ends are free to move, joints may be placed at 30 feet on center. Provisions should also be made for expansion where roofing abuts restraints such as walls or curbs.

Today, seamed joints are usually made using mechanical seaming devices.

Flashing.   Flashings for formed-in-place metal roofing are much the same as indicated in this chapter for other applications, except that perimeter flashings are usually of the same material and finish as the roofing. Metal flashing on field-formed metal roofing, other than cap (counter) flashings, are often locked into the metal roofing using a lock-seam joint. Refer to "Where to Get More Information" at the end of this chapter for guidance about metal thickness for flashings.

Flashings for preformed roofing are too often ignored by designers and general contractors and left up to roofing or mechanical system contractors. Those entitites, unfortunately, are not always very knowledgeable about metal roofing systems. Sometimes the flashing that results leaks in the first rain. Because of the ribs, flutes, and other irregularities that characterize much preformed roofing, flashings should be made specifically for the particular roofing or should be flexible enough to accommodate the roofing's shape. Base and other edge flashings should be manufactured by the roofing panel manufacturer specifically for the use intended. There are several manufacturers of flexible flashings for penetrations through preformed metal roofing.

Roofing Specialties.   The discussion of roof specialties earlier in this chapter applies to specialties used with metal roofing.

Roofing specialties associated with formed-in-place metal roofing are usually formed from the same metal with the same finish as the associated metal roofing. Lead roofing and gutter linings, cornice covers, and base flashings associated with lead roofing, for example, are usually formed from 3-pound lead. Cap flashings associated with lead roofing may be formed from 2-1/2-pound lead. Refer to "Where to Get More Information" at the end of this chapter for guidance about other metal thickness for roof specialties associated with formed-in-place metal roofing.

Specialties for use with preformed metal roofing should be especially formed to fit the roofing pattern, from the same materials and with the same finish as the roofing. Some metal roofing manufacturers, for example, produce curbs specifically for use with their product. Using standard curbs, skylights, roof hatches, and other accessories and specialties with ribbed preformed roofing will almost always lead to leaks.

Flashing, Roof Specialty, and Sheet Metal Roofing Failures and What to Do about Them

Failed flashing and roof specialties probably are responsible for more roof leaks than any other source. Many such failures are preventable.

Reasons for Failure

Flashings, roof specialties, and metal roofing may fail because of poor design (see Fig. 8-10), improper installation, bad materials, contact with chemicals,

Figure 8-10 This downspout was struck by a car, but the real reason for this damage was placing the downspout in a location where it was sure to be struck. (*Photo by author.*)

weathering, and exposure. In addition, metals are often damaged by galvanic action.

Poor Design.   Design problems include those related to the supporting and underlying construction. Improper substrates can lead to failures even when the flashing, roof specialties, and roofing are properly designed, well-selected, and installed properly (see Fig. 8-11). The underlying structure must be stable, sound, and strong and permit proper fastening of the supported elements. Nailers must be properly installed in correct locations and configurations. Wood should be pressure preservative treated. Expansion and contraction must be accounted for.

Poor design also includes selecting the wrong material for a particular use. An example is selecting an elastomeric flashing that is sensitive to sunlight for use where it will be exposed. Selecting metals that will be in contact with dissimilar metals which are likely to experience galvanic corrosion is another example of improper materials selection. Using materials that will not bond to each other, or are incompatible for other reasons, is yet another example. An example that many people may not be aware of is using the wrong roofing cement. Plastic roofing cement is not the same as

Figure 8-11   When the underlying structure fails, the roofing is doomed. (*Photo by author.*)

flashing cement, even though they are both made from cut-back asphalt. Both types have fibers added, but flashing cement has more and longer fibers, which make it stiffer than roofing cement. Flashing cement should be specified for vertical applications and cants, while roofing cement is adequate for horizontal applications such as the gravel stop flashing.

Poor design also includes selecting an inferior material. Poor quality will always lead to failure, regardless of how well the installation is made.

A similar problem is selecting a material that is quite acceptable for other installations but not desirable for the case at hand. An example is using galvanized metal flashing for a slate roof, where a better selection is a longer-lasting flashing, such as copper. Some plastic flashings, which work well with some elastomeric roofings, are incompatible with bituminous roofing.

A larger problem of that type is using metal flashing that, to exclude water, must be fastened directly to built-up bituminous, modified bituminous, or single-ply roofs. Differential movement between metal and those other materials makes failure certain. Not only does the metal expand much more than those roofing materials, the cements used to make the connections and the roofing membranes themselves shrink from three to five times as much as the metal in cold weather. The result is inevitably a split joint. The only question is how long the flashing will stay in service before the membrane separates from the flashing enough to allow the joint to leak.

It is not always possible to avoid using metal flashings or specialties with nonmetal roofing. Gravel stops on a built-up bituminous roof are an example. In such cases, the magnitude of the problem can be reduced by properly designing the installation. The gravel stop should never be placed beneath the roofing membrane. It should be, instead, installed after the membrane is in place, in a bed of roofing cement, and covered with at least two layers of felt flashing membrane.

Location of flashed conditions is also a design consideration. When possible, flashings should be kept away from locations where water might pond. An exception, of course, is roof drain flashing in low-slope roofs. When possible, ponding of water on roofs should be prevented altogether. In any case, flashings should be raised far enough above the roof surface so that ponded water or melting snow cannot bypass the flashing. Base flashings should extend at least 8 inches higher than the water will ever reach. The common height of 8 inches above the roofing may be insufficient when water will pond or snow will build up on the roof.

Many, perhaps most, flashing, roof specialty, and metal roofing problems are the direct result of poor detailing. Too often, architect's installation details are drawn without proper knowledge or study of industry-recognized standards or the recommendations of the product's manufacturer. Refer to "Where to Get More Information" at the end of this chapter for a listing of sources of good flashing, roof specialty, and metal roofing details.

A major detailing problem is stopping base flashing too low or extending it too high up a vertical surface. It takes at least 6 inches of vertical dimension to permit proper application. Base flashing higher than 12 inches will often sag as a result of gravity. When it is necessary to protect the wall, a better solution is using a properly supported metal wall covering.

Flashings must be designed to permit differential movement across the joint being flashed (see Fig. 8-12). Details should not require that base flashings be nailed to both roof and wall, for example. Cap flashing should be anchored to the wall, but not the base flashing. There may be no differential movement where the walls and deck are integral, but in every other case, walls will move relative to roofs. Some of the movement will be short-term because of temperature changes. Other movement, however, may be long-term or even permanent. Masonry walls may permanently change in dimension as they absorb or release moisture, for example. Flashing should be designed to permit such movement.

Specifying proper fastening of flashings is essential. Fastener type, size,

Figure 8-12 A leak waiting to happen. The problem here is failure to take into account differential movement between the masonry wall and the wood-framed overhang. (*Photo by author.*)

spacing, and staggering is specified in virtually every source of data about flashings.

Bitumen flowing beneath gravel stops or roofing edge strips occurs because of faulty design or improper installation.

Improper Installation. Proper installation means following the recommendation of manufacturers and appropriate industry standards-setting bodies, most of which are listed in this book's Bibliography.

Sometimes, simple things can lead to failure. Roofing cement applied in too thick a coating, for example, can crack as a result of shrinkage and admit water.

Bad Materials. Improperly manufactured metal sheets are not impossible but are rare. Finishes may come to a project damaged, but that should be corrected before they are installed. Inferior finishes may not be apparent until after installation, but they are usually covered under the manufacturer's warranty. The only solution to problems that stem from bad materials is to remove the materials and install good materials.

More common are poorly manufactured assemblies. Assemblies with joints that should be welded but are not is an example. Other problems, such as units out of square, or units with lines and arises not smooth or straight, also occur. Poorly manufactured assemblies appear in existing buildings sometimes, when previously undetected imperfections are brought to light by field conditions such as expansion, contraction, or an applied load. In either case, there are two solutions: repair the damaged item or return it for a new, properly made item.

Chemical Attack. Metals and many elastomeric materials are attacked by acids, alkalies, carbonates, fluorides, and other chemicals used in the construction and maintenance of buildings. Acids used in masonry cleaning are a particularly odious problem. Wet alkaline materials, such as mortar, plaster, and concrete, will etch aluminum. Unfortunately, there is not universal agreement in the construction industry about the effect of the chemicals in some materials on metals. Some guides, for example, say that the tannic acid in red cedar is not a problem with copper, while others disagree. The Red Cedar Shingle and Handsplit Shake Bureau recommends using either aluminum or galvanized steel flashing in valleys of red cedar shingle or shake roofs. A table in Ramsey/Sleeper's 1981 *Architectural Graphic Standards,* on the other hand, indicates that when either aluminum or zinc is in contact with red cedar, galvanic action will occur, implying that those materials should not be used to flash red cedar roofing.

Metals can be protected from chemical attack by preventing their contact with offending materials. Surfaces in contact with masonry, concrete, or mortar, for example, can be coated with alkali-resistant coatings such as heavy-bodied bituminous paint.

Weathering and Exposure.   Wind-driven materials, such as dirt and hail, can erode the surface of metals. Ice crystals may damage some surfaces, too. But much damage people assume to be the result of weathering and exposure to normal conditions is actually due to attack by chemicals in the air. Metals in industrial areas weather more quickly than those in rural areas (see Fig. 8-13). Corrosion is worst in seacoast areas, however. Aluminum will corrode rather quickly, even in rural areas, that lie near the coast, because of the saline content of the air. The effect is enhanced in southern areas where the sun is hotter.

Some elastomeric materials are sensitive to the ultraviolet component of sunlight and will deteriorate rapidly when exposed to it.

Galvanic Corrosion.   Galvanic corrosion (electrolysis) happens when dissimilar metals are each contacted by an electrolyte, such as water. The

Figure 8-13 Poor maintenance contributed to the early deterioration of this low-cost metal roofing installation. (*Photo by author.*)

process is like that which takes place inside an automobile battery. The electrolyte causes an electric current to flow between the two metals, which removes material from one metal and deposits it on the other.

The electrolyte can be rainwater or just condensation. The effect is more severe when the dissimilar materials are close together, but even wash-off from one material can corrode a dissimilar material. Rainwater washing across copper flashing can destroy an aluminum gutter, for example.

Some combinations of materials are more susceptible to galvanic corrosion than others. Galvanic tables have been developed which show which materials are likely to create galvanic corrosion. The propensity for galvanic action depends on where the two materials fall in the galvanic table. The farther apart two materials are in the table, the less severe will be the corrosion. For example, aluminum, which falls near one end of most tables, will corrode when allowed to become wet while in contact with lead, brass, bronze, iron, or steel. When dissimilar materials are attacked by galvanic action, it is the material that is highest in the galvanic table which will corrode.

The list of metal materials earlier in this chapter is arranged roughly in galvanic table order. It should be used carefully, however, because the order is only an approximation gleaned from available references, which do not totally agree with each other. Neither do sources which do not include tables agree about which materials are compatible. Inconsistent data are only marginally useful when one is trying to determine the cause of existing corrosion. Perhaps someday, some organization will study galvanic action in sufficient detail to resolve the conflicts that now exist. Meanwhile, probably the best way to resolve a corrosion problem that persists after other possible causes have been eliminated is to assume that the culprit is the adjacent material, regardless of what a published galvanic series table, or any other source, says.

There are two ways to prevent corrosion due to galvanic action. One is to avoid using dissimilar metals in close contact. When that is not possible, it is necessary to prevent an electrolyte from contacting both. A heavy-bodied bituminous paint is sometimes used to coat one of the surfaces. Aluminum paint can be applied to dissimilar surfaces to protect aluminum. The metals can be separated using moisture-resistant building felts. Other separation means are also used.

Natural Deteriorations.   Some deterioration, of course, is the result of aging. Every material has a natural life span. At the end of the span, materials will begin to fail, especially if they have been neglected (see Fig. 8-14). Plastics will harden and crack. Metals will fatigue from normal flexing and split. When failure is due to aging, the only solution is to remove the material and rebuild the flashing using new materials.

Figure 8-14   Neglected galvanized metals will eventually rust away. (*Photo by author.*)

Evidence of Failure

Inspection.   Roof flashings, roof specialties, and metal roofing should be inspected regularly (see Fig. 8-15). Refer to Chapter 6, "Evidence of Failure," subparagraph "Inspection," for a discussion of the process of finding a leak in a steep roof and to the similar paragraph in Chapter 7 for a discussion of finding leaks in low-slope roofs.

When the leak is in a flashing, examine all flashings to ensure that other conditions do not exist which are likely to leak soon.

All flashing and roof specialties should be examined by a knowledgeable inspector at least once a year (some experts say every 6 months). Particular attention should be paid to conditions where metal is attached directly to, or capped by, bituminous, modified bitumen, or elastomeric membrane roofing, flashings, or stripping to ensure that the membrane is not separating from the metal. The drainage system should be inspected for damage (see Fig. 8-16).

Cap flashings that are directly fastened to vertical surface without use of a reglet or other proper receiver, even where a clamping bar is used,

Figure 8-15 A ground level walk-by is all the inspection necessary to discover this broken gutter joint. The dark material is mildew. (*Photo by author.*)

should be inspected at least once every 6 months to ensure that the fasteners are not loose and the sealant at the top of the flashing has not dried out.

Pitch pans should also be inspected at least once every 6 months and repaired at the slightest sign of alligatoring or splitting of the fill material or deterioration of the metal pan.

Breaks and Holes. Many flashing failures appear as breaks or tears in the flashing itself. Probably the most common cause for such damage is people walking or working on the roof.

Differential movement between substrates on opposite sides of a flashed joint can also cause breaks in the flashing or in soldered joints between flashing sheets.

Figure 8-16 Neglected damage can lead to unsightly and even unsanitary conditions, as this fungus growth demonstrates. (*Photo by author.*)

Separation.   A major cause of leaks at flashings is separation of the flashing from roofing, stripping, other flashing, and walls.

The most common cause of such separations is differential thermal movement between the flashing and adjacent materials. Such separation is not usually caused by metal shrinkage, as is often assumed, but rather by the membrane shrinking away from the metal.

Separation can also be caused by differential movement in the substrates, when the flashing bridges the joint and is not designed to accommodate the movement.

Another potential for separation is flashings, such as vent pipe and equipment support flashings, that are supported on the roofing membrane rather than on the deck. Such flashings sometimes lack stability and tend to move relative to the membrane more than structures fastened to the deck.

Leaks around or through Flashings.   Leaks can occur even when there are no breaks in the flashing if water is permitted to bypass the flashing. Such bypassing can occur when the flashing is poorly designed, such as when the top of base flashing is too low, or when the lap between cap and base flashing is too small. Ponding is a major cause of leaks around sound flashing.

An easily solved leak around flashing is due to lack of or failure of the sealant in reglets.

Water can penetrate directly through some kinds of flashings when they are installed too thin or in two few layers. Water leaks directly through sound metal flashings are not possible.

Finish Failures.   Paint failures occur because of improper selection of paint material or primer, improper application, or age. Paint may blister, alligator, crack, or peel (see Fig. 8-17).

Zinc coatings may be damaged, permitting the underlying metal to rust.

Factory coatings may prematurely fade, delaminate, blister, alligator, crack, or peel.

Figure 8-17   Paint failure on a metal roof. (*Photo by author.*)

Repairing Existing Sheet Metal Work—General Requirements

Compatibility.   To prevent galvanic corrosion, patches should be the same metal as the metal being patched. Paint or other coatings, or membranes, can sometimes be used to prevent contact between a dissimilar patch and the metal patched, but not often, because interposed coatings preclude soldering or welding. In addition, where the surface being patched is ribbed or patterned, coatings or membranes can sometimes prevent the patch from sealing. Where using the same metal is not practicable, patches should be made from a metal known not to cause corrosion when in contact with the surface being patched. Where there is a question about a desired patch metal's position in the galvanic table relative to the metal to be patched, select a different metal, the position of which is absolutely known. The most reliable guide in such instances is personal experience. When personal experience with the two metals in question does not resolve the issue, consult both metals' producers. It is not a good idea to rely on third-party sources of information about position in galvanic tables.

Compatibility between a patch and the metal to be repaired will also increase the number of potential installation methods. Dissimilar metals probably cannot be soldered or welded.

Incompatibility can be implicated in separations due to differential thermal movements. Preventing such implication is easy. Do not use incompatible materials, and do not try to fasten metal to bituminous or elastomeric materials. A cure is almost impossible, except by removing the metal and substituting a material that is compatible with the adjacent materials.

Weather Conditions.   A common cause of roofing system failure is water trapped beneath or between membranes or between flashings and membranes. Sometimes, the trapped water comes from adjacent construction, such as masonry parapet walls. Often, however, the moisture is built in as the roofing and flashings are installed. It is essential, when repairing flashings, to first dry out the substrates and adjacent membranes to eliminate all trapped moisture. It is, therefore, necessary to repair flashings only during dry weather and not before precipitation or condensation has dried.

Temporary Fixes.   It is sometimes possible to temporarily repair minor flashing and metal roofing leaks with a patch of bituminous roofing cement or flashing cement. Such repairs are always temporary, whether planned that way or not.

Removing Existing Sheet Metal versus Patching.   Leaving existing metal flashing in place when new flashing is to be installed is generally not a good idea. Often it is not possible. Sometimes, however, such as when roofing over existing steep roofing, metal flashing may be left in place.

In most cases, damaged metal flashing should be either patched or removed and new flashing installed. Determining whether to patch or remove the existing flashing depends on the answers to the following questions.

Is it possible to satisfactorily repair the existing flashing? It might be necessary to have a qualified roofer, architect, roofing consultant, or the flashing manufacturer's representative look at the condition in the field before making this determination.

Will the cost of repairing equal or exceed the cost of new flashing? If so, the new flashing is the best solution.

Is the existing flashing condition well designed, or does bad design contribute to the problem?

Permanent Repairs.   When faulty design is responsible for a flashing failure, it is necessary to correct the design error before repairing the flashing. Correcting design errors can be costly, but if they are allowed to remain, leak repairs will not be permanent. For example, a leaking flashing which is below ponded water or snow buildup will probably continue to leak, regardless of the remedial measures taken (see Fig. 8-18). There, it is necessary to either eliminate the ponding or snow buildup or raise the level of the flashing.

Similarly, damage due to continuing differential movement will recur unless the movement is stopped (usually not possible) or the flashing is redesigned to accommodate the movement. Resoldering joints that have split because of differential movement, for example, will hold only temporarily unless the movement is prevented from affecting the joint. Sometimes, an expansion joint must be installed between the two elements that are moving differentially. When the flashing between a roof and a parapet wall is opening because of differential movement, for example, it is sometimes necessary to install a curb an inch or so away from the wall to receive the base flashing. The curb is then capped with a metal cap designed to permit movement and fastened only to the wall (see Fig. 7-1).

Leaks at roof penetrations will also recur after repairs if the flashing design is faulty. It would be better to build a curb around most penetrations, anyway, regardless of their size. When the flashing at a penetrating item has leaked and the existing flashing defies satisfactory repair, a curb becomes essential. It does no good to continually repair a flashing that is poorly designed. If a curb is already present, but is leaking, it may be necessary to install structural stiffening devices to more firmly support the curb. It may not be possible to stop a curb of the wrong shape on preformed metal

Figure 8-18 Mildew stains on these walls resulted from ponded water pouring over the roof edge. Repairing the broken gutter and installing a new downspout will not prevent this damage until the overflow is prevented from recurring. (*Photo by author.*)

roofing from leaking. It may be necessary to remove the curb and install a new curb of the proper configuration for that roofing.

When rooftop equipment supports prove impossible to satisfactorily flash, it may be necessary to redesign the supports.

Improperly installed flashings should be removed and, if in satisfactory condition, reinstalled properly. Patching or wedging loose flashings in place where existing flashings are improperly installed usually yields unsatisfactory results.

Whatever the reason for the failure, flashings having broken seams, holes, splits, or tears in the flashing material, or a severely corroded surface should be removed and new flashing installed. At every flashing location, loose stripping felts should be removed and new felts installed.

Leaks through unsealed reglets can be easily repaired by filling the reglet with sealant.

Other leaks at cap flashings may require removal of the flashing. New cap flashing installed to replace a damaged section should be lapped beneath the cap flashing left in place by at least 3 inches.

Where reglets are missing, improperly installed, or of a type that has failed, it may be possible to install the cap flashing in raggles (cut slots) in the wall. Raggles should be 1-1/2 inches deep. Cap flashings should be secured in raggles or reglets using lead or folded metal wedges spaced about 24 inches apart. Reglets and raggles should then be sealed with elastomeric sealant to make the joint watertight.

Cap flashing which has been nailed directly to the wall, even when a clamping bar is used, often fails. Sometimes the failure is due to the sealant at the top of the flashing drying out. Sometimes the sealant will have been omitted altogether. Sometimes, such flashings will fail when the fasteners come loose in the joints, permitting the flashing to fall away from the wall. It might be possible to repair such conditions by refastening the flashing and resealing the joint at the top using an elastomeric sealant. Probably the best solution to such flashing failures, however, is to remove the failed flashing and reinstall new flashing in a reglet or raggle.

Leaking metal edge strips and gravel stops with open or broken joints should be removed (see Fig. 8-19). Even when the metal is not damaged,

Figure 8-19   A broken gravel stop soldered joint caused this staining. (*Photo by author.*)

stripping that flashes the edge strip or gravel stop into the roofing membrane which has separated from the metal should be removed completely. When the stripping has delaminated or the edge strip or gravel stop is damaged and the edge strip or gravel stop has been installed at the level of the roofing, it is worthwhile to consider rebuilding the edge condition. Installing metal edge strips or gravel stops flush with the roof often contributes to flashing failures, because they sit in small puddles of water after every rain. Before reinstalling the roof edge strip or gravel stop, install a nailer and tapered cant to elevate the metal above the standing water level. Provide scuppers to permit water to drain off the roof if there are no interior drains.

Removed stripping should be discarded. Removed edge strips and gravel stops should be discarded if in poor condition but may be reinstalled if in good condition. The new or reinstalled edge strip or gravel stop should be set in roofing cement and stripped in with two layers of the type of membrane recommended by the roofing manufacturer.

On some roofs, both roof edge strips and gravel stops have been omitted and the roofing bent over the edge of the roof and fastened (or not fastened) to the fascia. Proper repair of such conditions requires installation of metal edge strips or gravel stops.

Bitumen flowing beneath roof edging or gravel stops can only be corrected by removing the gravel stop or edge strip and rebuilding the roofing to provide proper bitumen stops. Unless the situation is particularly bad, or the gravel stop or edge must be removed for other reasons, this is probably not worth the effort.

Formed metal roof expansion joints that are flush with the roof or which have metal flanges flashed into membrane roofing are a common source of leaks. Such conditions should not be repaired, however. The only proper solution to such conditions is to remove the offending flashing and install a proper expansion joint on curbs. Some experts believe that the same applies to an expansion joint cover made of elastic materials. Any joint meant to move, they say, does not belong at roof membrane level.

Pitch pans are by nature faulty designs. When the filler material inevitably cracks and admits water, removing the filler and pouring new material into the pan will often stop the leak, until the next time it fails. Pitch pans are a continuing source of trouble. It would be better to remove the pitch pan and install a properly constructed curb and cap flashing. There are, unfortunately, some coditions that are almost impossible to flash other than with a pitch pan. In those cases, there is no choice but to periodically renew the fill material.

Breaks, splits, and holes in metals that can be soldered should usually be repaired by soldering on a patch of the same material as that being patched. Galvanized steel, copper, and terne-coated stainless steel can be soldered. Patches in lead can be fastened by welding. Those metals that can be soldered can be soldered even when they are prefinished. Unfortunately, to solder any metal, it is necessary to remove all coatings and finishes. Field-finished materials can be easily refinished. Refinishing factory-finished materials to match the existing ones, however, is not easy. It is possible to field paint the patches, of course, but doing so will not often produce a finish and color that looks like the adjacent factory-finished materials.

Damaged field-applied finishes, such as paint or aluminum coatings, can be repaired by refinishing. Rust, dirt, and foreign materials should be removed before refinishing begins.

The best solution to repairing factory-finished metals is to remove the damaged piece and replace it with a new piece of the same material as was originally used. Even the same material may not match because of aging of the original finish in the field. New prefinished materials can be installed only using mechanical means such as fasteners.

Metals that cannot be soldered, such as aluminum and materials with bituminous coatings, can be often patched by covering the damaged area with a glass fiber reinforced sheet set in roofing or flashing cement. If the material is prefinished, the same advice given in the preceding paragraph applies.

It is often not possible to repair metal roofing properly using flat sheet materials, even when the roofing is field-finished, because of the diversity of sheet patterns used (see Fig. 8-20).

Repairing, Replacing, and Re-Covering Metal Roofing

In addition to the information in this chapter, refer to the following portions of Chapter 6 for additional guidance about metal roofing.

Considerations when deciding whether to remove existing roofing or leave existing roofing in place and roof over it: Refer to "Repair, Replacement, and Re-Covering Work—General Requirements," and "Removing Existing Roofing vs. Roofing Over" in Chapter 6.

General requirements for preparing existing roofing to receive new roofing: "Preparing Existing Roofing to Receive New Roofing." Most general references and specific references to metal tile in Chapter 6 apply to metal roofing.

Using existing underlayment: "Preparing Existing Underlayment to Receive New Roofing."

In addition, the recommendations in "Repairing Metal Tile Roofing" and "Installing New Metal Tile Roofing over Existing Materials" generally apply to metal roofing of the types discussed in this chapter. In most cases, where "metal tile" appears in Chapter 6, substitute "metal roofing."

Figure 8-20 A poorly made and unsightly patch on a metal roof. Part of an earlier failed patch can be seen immediately above the new patch. This owner must find a knowledgeable roofer who is also a competent craftsperson. (*Photo by author.*)

Where to Get More Information

The Brick Institute of America's 1985 publication *Technical Notes on Brick Construction,* 7A, Water Resistance of Brick Materials, Part II or III, Includes BIA's recommendations about flashings for brick walls.

C. W. Griffin's 1982 book *Manual of Built-Up Roof Systems* includes a comprehensive discussion of the principles of roof flashing design for low-slope roofs. It also includes some flashing details to demonstrate the points made in the discussion, and an excellent review of the causes of flashing failure. Many details from an earlier edition of the National Roofing Contractors Association's *The NRCA Roofing and Waterproofing Manual* are also included. Anyone with a flashing or roofing problem should obtain a copy of Mr. Griffin's book.

The following publications include discussion of structural metal roofing, which is excluded from this text:

Paul D. Nimtz's 1987 article "Comparing Structural and Architectural Metal Roofing."

Thomas R. Scharfe's 1987 article "Owners, Designers, Discover Benefits of Metal Roofs."

Burt Shell's 1988 article "Standing Seam Roof Continues Inrodes in Educational Market."

Fred Stephenson's 1986 article "Clip Design Critical in Standing Seam Roofing."

The following publications include good details for roof flashing and roof specialties for both low-slope and steep roofs and include requirements for metal thickness for most conditions:

American Iron and Steel Institute's (AISI) *Stainless Steel: Suggested Practices for Roofing, Flashing, Copings, Fascias, Gravel Stops, and Drainage.*

The Copper Development Association's 1980 pamphlet *Sheet Copper Applications.*

The National Roofing Contractors Association's 1986 *The NRCA Roofing and Waterproofing Manual.*

The Ramsey/Sleeper 1981 *Architectural Graphic Standards.*

Sheet Metal and Air-Conditioning Contractors National Association's (SMACNA) *Architectural Sheet Metal Manual.*

Publications of the following organizations include details and recommendations for flashing and roof specialties related to the products of their members:

National Slate Association, particularly in the 1926 book *Slate Roofs.*

Red Cedar Shingle and Handsplit Shake Bureau.

The following publications include good metal roofing details:

American Iron and Steel Institute's (AISI) *Stainless Steel: Suggested Practices for Roofing, Flashing, Copings, Fascias, Gravel Stops, and Drainage.*

The Copper Development Association's 1980 pamphlet *Sheet Copper Applications.*

The Lead Industries Association's *Lead Roofing and Flashing.*

The Ramsey/Sleeper 1981 *Architectural Graphic Standards.*

Sheet Metal and Air-Conditioning Contractors National Association's (SMACNA) *Architectural Sheet Metal Manual.*

In addition, the following publications offer good details and recommendations for specific flashing and roof specialty conditions:

Paul Heineman, in his article "Coping With Membrane Roof Penetrations," in the November 1987 *The Construction Specifier,* provides a great deal of information and some details for penetrations through low-slope roofs and for roof mounted equipment on low-slope roofs.

Sheldon B. Israel's "The Nuts and Bolts of Peripheral Roofing Components" in the November 1987 issue of *The Construction Specifier* includes some excellent and helpful recommendations about roof penetrations and specialties.

The Air-Conditioning and Refrigeration Institute (ARI), Sheet Metal and Air-Conditioning Contractors National Association, and National Roofing Contractors Association joint publication *Guidelines for Roof Mounted Outdoor Air-Conditioner Installations,* includes details for duct, piping, and conduit penetrations through roofing.

The National Roofing Contractors Association and Asphalt Roofing Manufacturers Association 1988 publication *NRCA/ARMA Manual of Roof Maintenance and Repair* is an excellent source of information for dealing with existing flashing conditions on low-slope roofs. This booklet is a must buy for anyone who owns a low-slope roof.

In addition to the sources of details listed above, many manufacturers offer suggested details for application of their products. Some even offer repair recommendations.

Data Sources

NOTE: The following list includes sources of data referenced in the text, included in the Bibliography, or both. **HP** following a source indicates that that source also contains data of interest to those concerned with historic preservation.

Adhesive and Sealant Council (ASC)
1500 Wilson Boulevard, Suite 515
Arlington, VA 22209
Tel: (703) 841-1112

Advisory Council on Historic
    Preservation
1100 Pennsylvania Avenue, Suite 809
Washington, DC 20004
Tel: (202) 786-0503 **HP**

AIA Service Corporation
The American Institute of Architects
1735 New York Avenue, N.W.
Washington, DC 20006
Tel: (202) 626-7300

Aluminum Association (AA)
818 Connecticut Ave., N.W.
Washington, DC 20006
Tel: (202) 862-5100

American Architectural Manufactur-
    er's Association (AAMA)
2700 River Road, Suite 118
Des Plaines, IL 60018
Tel: (312) 699-7310

American Association of State and
    Local History
172 Second Ave, North, Suite 102
Nashville, TN 37201
Tel: (615) 255-2971 **HP**

American Council of Independent
    Laboratories (ACIL)
1725 K Street N.W.
Washington, DC 20006
Tel: (202) 887-5872

The American Institute of Architects
1735 New York Avenue, N.W.
Washington, DC 20006
Tel: (202) 626-7300

American Institute of Architects
Committee on Historic Resources
1735 New York Ave, N.W.
Washington, DC 20006
Tel: (202) 626-7300 **HP**

American Iron and Steel Institute
1133 15th Street, Suite 300
Washington, DC 20005
Tel: (202) 452-7100

American National Standards Institute (ANSI)
1430 Broadway
New York, NY 10018
Tel: (212) 354-3300

American Plywood Association (APA)
P.O. Box 11700
Tacoma, WA 98411
Tel: (206) 565-6600

American Society of Heating, Refrigerating and Air-Conditioning Engineers, Inc.
1791 Tullie Circle, N.E.
Atlanta, GA 20329
Tel: (404) 636-8400

Asphalt Institute
Asphalt Institute Building
College Park, MD 20740
Tel: (301) 277-4258

Asphalt Roofing Manufacturers Association (ARMA)
6288 Montrose Road
Rockville, MD 20852
Tel: (301) 231-9050

Association for Preservation Technology
Box 2487 Station D
Ottawa, ONT KIP 5W6, Canada
Tel: (613) 238-1972 **HP**

ASTM
1916 Race Street
Philadelphia, PA 19103-1187
Tel: 215-299-5585 Fax: 215-977-9679
TWX 710-670-1037

Brick Institute of America
11490 Commerce Park Drive, Suite 300
Reston, VA 22091
Tel: (703) 620-0010

Building Design and Construction
Cahners Plaza
1350 East Touhy Avenue

P.O. Box 5080
Des Plaines, IL 60018
Tel: (312) 635-8800

Building Stone Institute
420 Lexington Avenue
New York, NY 10017
Tel: (212) 490-2530
Note: BSI has no technical personnel. Author recommends contacting individual producers associations or producers rather than BSI.

Building Thermal Envelope Coordinating Council
1015 15th Street, N.W.
Washington, DC 20005

Certified Ballast Manufacturers Association (CBM)
Hanna Building, Suite 772
1422 Euclid Avenue
Cleveland, OH 44115
Tel: (216) 241-0711

The Cold Regions Research and Engineering Laboratory
72 Lyme Road
Hanover, NJ 03755
Tel: (606) 646-4200

The Construction Specifier
The Construction Specifications Institute
601 Madison Street
Alexandria, VA 22314-1791
Tel: (703) 684-0300

Copper Development Association, Inc.
Greenwich Office Park 2
Box 1840
Greenwich, CT 06836
Tel: (203) 625-8210

Council of Forest Industries of British Columbia
1055 West Hastings Street
Vancouver, B.C. V6E 2H1

Edgell Communications, Inc.
7500 Old Oak Boulevard
Cleveland, OH 44130

Exteriors
1 East First Street   .
Duluth, MN 55802
Tel: (218) 723-9200

Factory Mutual System
1151 Boston-Providence Turnpike
Norwood, MA 02062
Tel: (617) 762-4300

General Services Administration
General Services Building
18th and F Streets, N.W.
Washington, DC 20405
Tel: (202) 655-4000

Hartford Architecture Conservancy
130 Washington Street
Hartford, CT 06106
Tel: (203) 525-0279 **HP**

Heritage Canada Foundation
Box 1358, Station B
Ottowa, ONT KIP 5R4, Canada
Tel: (613) 237-1867 **HP**

Historic Preservation (Magazine)
(See National Trust for Historic Pres-
    ervation) **HP**

Illinois Historic Preservation Agency
Division of Preservation Services
Old State Capitol
Springfield, IL 62701
Tel: (217) 782-4836 **HP**

Indiana Limestone Institute of
    America, Inc.
Stone City Bank Building, Suite 400
Bedford, IN 47421
Tel: (812) 275-4426

Institute for Applied Technol-
    ogy/Center for Building
    Technology
National Bureau of Standards
U.S. Department of Commerce

Washington, DC 20540
Tel: (202) 342-2241

International Masonry Institute
823 15th Street, N.W. #1001
Washington, DC 20005
Tel: (202) 783-3908

The Lead Industries Association
292 Madison Avenue
New York, NY 10017
Tel: (212) 578-4750

Library of Congress
First Street, N.E.
Washington, DC 20540
Tel: (202) 287-5000 **HP**

Marble Institute of America
33505 State Street
Farmington, MI 48024
Tel: (313) 476-5558

McGraw Hill Book Company
1221 Avenue of the Americas
New York, NY 10020
Tel: (212) 997-2271

Metal Architecture
123 North Polar Street
Fostoria, OH 44830
Tel: (419) 435-8571

National Alliance of Preservation
    Commissions
Hall of the States
444 N. Capitol St., N.W., Suite 332
Washington, DC 20001
Tel: (202) 624-5490 **HP**

National Building Granite Quarries
    Association
See latest Sweets Catalog Files.
    Representative rotates among
    association members

National Bureau of Standards (NBS)
See National Institute of Standards
    and Technology.

National Institute of Standards and
    Technology

(Formerly National Bureau of
   Standards)
Gaithersburg, MD 20234
Tel: (301) 975-2000

National Institute of Standards and
   Technology
Center for Building Technology
Gaithersburg, MD 20234
Tel: (301) 975-5900

National Concrete Masonry
   Association
P.O. Box 781
Herndon, VA 22070
Tel: (703) 435-4900

National Preservation Institute
P.O. Box 1702
Alexandria, VA 22313
Tel: (703) 393-0038 **HP**

National Roofing Contractors Associ-
   ation (NRCA)
One O'Hare Centre, 6250 River Road
Rosemont, IL 60018
Tel: (312) 318-6722

National Tile Roofing Manufacturers
   Association, Inc.
c/o W. F. Pruter Associates
3217 Los Feliz Boulevard
Los Angeles, CA 90039
Tel: (213) 660-6259

National Trust for Historic
   Preservation
1785 Massachusetts Avenue, N.W.
Washington, DC 20036
Tel: (202) 673-4000 **HP**

The Old House Journal
69A Seventh Ave.
Brooklyn, NY 11217
Tel: (718) 636-4514

Portland Cement Association
5420 Old Orchard Road
Skokie, IL 60077
Tel: (312) 966-6200

The Preservation Press
National Trust for Historic
   Preservation
1785 Massachusetts Avenue, N.W.
Washington, DC 20036
Tel: (202) 673-4000 **HP**

Preservation Resource Group
5619 Southampton Drive
Springfield, VA 22151
Tel: (703) 323-1407 **HP**

Professional Roofing
(Formerly *Roofing Spec*)
Contact National Roofing Contractors
   Association

Red Cedar Shingle and Handsplit
   Shake Bureau
515 116th Avenue, N.E., No. 275
Bellevue, WA 98004
Tel: (206) 453-1323

Roof Consultants Institute
7424 Chapel Hill Road
Raleigh, NC 27607
Tel: (919) 859-0742

The Roofing Industry Education Institute (RIEI)
7006 South Alton Way, No. B
Englewood, CO 80112
Tel: (303) 770-0613

Roofing '86
See National Roofing Contractors
   Association.

Roofing Spec
See Professional Roofing.

Rubber Manufacturer's Association
1400 K Street, N.W.
Washington, DC
Tel: (202) 682-4800

Sealant Engineering and Associated
   Lines Association (SEAL)
P.O. Box 24302
San Diego, CA 92124
Tel: (619) 569-7906

Sealant and Waterproofing Institute
3101 Broadway, Number 300
Kansas City, MO 64111-2416
Tel: (816) 561-8230

Sheet Metal and Air-Conditioning
    Contractors National Associa-
    tion, Inc. (SMACNA)
8224 Old Courthouse Road
Vienna, VA 22180
Tel: (703) 790-9890
    For publications contact:
SMACNA Publications Department
P.O. Box 70
Merrifield, VA 22116
Tel: (703) 790-9890

Single Ply Roofing Institute (SPRI)
104 Wilmot Road, Suite 201
Deerfield IL 60015-5195
Tel: (312) 940-8800

Society for the Preservation of New
    England Antiquities
141 Cambridge Street,
Boston, MA 02114
Tel: (617) 227-3956 **HP**

Thermal Insulation Manufacturers
    Association (TIMA)
7 Kirby Plaza
Mt. Kisco, NY 10549
Tel: (914) 241-2284

Transportation Research Board
National Academy of Sciences

2101 Constitution Avenue, N.W.
Washington, DC 20418
Tel: (202) 334-2000 or 334-2222

Technical Preservation Services
U.S. Department of the Interior,
Preservation Assistance Division
National Park Service
Washington, DC 20013-7127 **HP**

Underwriters Laboratories
333 Pfingsten Road
Northbrook, IL 60062
Tel: (321) 272-8800

U.S. Department of the Interior
National Park Service
P.O. Box 37127
Washington, DC 20013-7127
Tel: (202) 343-7394 **HP**

U.S. General Services Administration
Historic Preservation Office
Washington, DC 20405
Tel: (202) 655-4000 **HP**

Van Nostrand Reinhold
115 Fifth Avenue
New York, NY 10003
Tel: (212) 254-3232

John Wiley and Sons
605 Third Avenue
New York, NY 10158
Tel: (212) 850-6000

Bibliography

Note: Each item in the Bibliography is followed by one or more numbers in brackets. Those numbers list the chapters in this book to which that Bibliographical entry applies.

The **HP** following some entries in the Bibliography indicates that that entry has particular significance for historic preservation projects.

Sources for many of the entries, including addresses and telephone numbers, are listed in the Appendix.

AIA Service Corporation. *Masterspec,* Basic: Section 07110, Bituminous Waterproofing, 5/85 Edition. The American Institute of Architects. [4]

——— *Masterspec,* Basic: Section 07115, Sheet Waterproofing, 8/87 Edition. The American Institute of Architects. [4]

——— *Masterspec,* Basic: Section 07120, Fluid-Applied Waterproofing, 5/85 Edition. The American Institute of Architects. [4]

——— *Masterspec,* Basic: Section 07160, Bituminous Dampproofing, 5/85 Edition. The American Institute of Architects. [4]

——— *Masterspec,* Basic: Section 07175, Water Repellents, 8/85 Edition. The American Institute of Architects. [3]

——— *Masterspec,* Basic: Section 07200 - Insulation, 11/85 Edition. The American Institute of Architects. [2, 5]

——— *Masterspec,* Basic: Section 07311, Asphalt Shingles, 11/85 Edition. The American Institute of Architects. [6, 8]

——— *Masterspec,* Basic: Section 07317, Wood Shingles and Shakes, 11/85 Edition. The American Institute of Architects. [6, 8]

——— *Masterspec,* Basic: Section 07410, Preformed Roofing and Siding, 8/83 Edition. The American Institute of Architects. [8]

—— *Masterspec,* Basic: Section 07510, Built-Up Asphalt Roofing System, 8/85 Edition. The American Institute of Architects. [2, 5, 7, 8]

—— *Masterspec,* Basic: Section 07530, Flexible Sheet Roofing System, 8/87 Edition. The American Institute of Architects. [2, 5, 7, 8]

—— *Masterspec,* Basic: Section 07600, Flashing and Sheet Metal, 2/84 Edition. The American Institute of Architects. [8]

—— *Masterspec,* Basic: Section 07620, Metal Fascias and Copings. 2/84 Edition. The American Institute of Architects. [8]

—— *Masterspec,* Basic: Section 07700, Roof Specialties and Accessories, 5/87 Edition. The American Institute of Architects. [8]

Air-Conditioning and Refrigeration Institute (ARI), Sheet Metal and Air-Conditioning Contractors National Association (SMACNA), and National Roofing Contractors Association (NRCA), *Guidelines for Roof Mounted Outdoor Air-Conditioner Installations.* ARI, SMACNA, and NRCA. [8]

Allen, Edward. 1985. *Fundamentals of Building Construction: Materials and Methods.* New York: Wiley. [6, 8]

American Iron and Steel Institute. 1972. *Stainless Steel: Suggested Practices for Roofing, Flashing, Copings, Fascias, Gravel Stops, and Drainage.* Washington, DC: AISI. [8]

American Society of Heating, Refrigerating and Air-Conditioning Engineers, Inc. 1985. *ASHRAE Handbook: 1985 Fundamentals.* Atlanta, GA: ASHRAE. [2, 5, 8]

Anderson, Brent. 1986. Waterproofing and the Design Professional. *The Construction Specifier,* March, 39(3):86–97. [4]

Ashurst, John. 1988. *Practical Building Conservation, Volume 2.* Aldershot, England: Gower. [2, 8]

Asphalt Roofing Manufacturers Association. 1987. *Built-Up Roofing Systems Design Guide.* Rockville, MD: Asphalt Roofing Manufacturers Association. [5, 7]

Asphalt Roofing Manufacturers Association (ARMA) and National Roofing Contractors Association (NRCA). 1988. *Quality Control Recommendations for Polymer Modified Bitumen Roofing.* Asphalt Roofing Manufacturers Association and National Roofing Contractors Association. This document was reprinted in the March 1988 magazine *Professional Roofing* published by NRCA. [5, 7]

Association for Preservation Technology. 1969. *Bulletins of APT, Volume 1.* Ottawa, Ontario: APT. **HP**

ASTM. Standard C-208. Specification for Insulating Board (Cellulosic Fiber), Structural and Decorative. ASTM. [5]

—— Standard C-270. Specification for Mortar for Unit Masonry. ASTM. [6]

—— Standard D-312. Specification for Asphalt Used in Roofing. ASTM. [7]

—— Standard D-449. Specifications for Asphalt Used in Dampproofing and Waterproofing. ASTM. [4]

——— Standard D-450. Specification for Coal-Tar Bitumen Used in Roofing, Damp-proofing, and Waterproofing. ASTM. [4, 7]

——— Standard D-491. Specification for Asphalt Mastic Used in Waterproofing. ASTM. [4]

——— Standard D-1187. Test Method for Asphalt Emulsions for Use as Protective Coatings for Metal. ASTM. [4]

——— Standard D-1653. Test Method for Moisture Vapor Permeability of Organic Coating Films. ASTM. [3]

——— Standard D-4433. Standard Specifications for Poly (Vinyl Chloride) Sheet Roofing. ASTM. [7]

——— Standard D-4637. Specifications for Vulcanized Rubber-Sheet Used in Single Ply Roofing Membranes. ASTM. [7]

——— Standard E-96. Test Method for Water Vapor Transmission of Materials. ASTM. [2, 3, 4, 5, 6, 7, 8]

——— Standard E-108. Fire Tests of Roof Coverings. ASTM. [5, 6, 7, 8]

——— Standard E-514. Test Method for Water Permeance of Masonry. ASTM. [3, 6]

Baker, Maxwell C. 1980. *Roofs: Design, Application and Maintenance* Montreal: Multiscience. (May be available from The National Roofing Contractors Association.) [6, 7, 8]

Baumgardner, Gaylorn. 1986. Modified Bitumens: Versatility in Reroofing. *Exteriors,* Autumn, 4(3):62–64. [7]

Baxter, Dick. 1986. 1001 Reasons Not to Roof over Wet Insulation. *Roofing Spec,* August, 14(8):27–30. [5, 7]

Bradford, Dane. 1987. BUR Repairs Made Easy with Torchable Modified Bitumen. *Roofing Spec,* March, 15(3):22–24. [7]

Brick Institute of America. 1980 (December). *Technical Notes on Brick Construction,* 7—Dampproofing and Waterproofing Masonry Walls. Brick Institute of America. [3, 4]

——— 1985. *Technical Notes on Brick Construction,* 7A—Water Resistance of Brick Materials, Part II of III. Brick Institute of America. [8]

——— 1965 (February). *Technical Notes on Brick Construction,* 7C—Moisture Control in Brick and Tile Walls: Condensation. Brick Institute of America. [2]

——— 1968 (June). *Technical Notes on Brick Construction,* 7D—Moisture Control in Brick and Tile Walls: Condensation Analysis. Brick Institute of America. [2]

——— 1976 (September/October). *Technical Notes on Brick Construction,* 7E—Colorless Coatings for Brick Masonry. Brick Institute of America. [3]

Brunnell, Gene. 1977. *Built to Last: A Handbook on Recycling Old Buildings.* Washington, DC: Preservation Press. **HP**

Bryant, Terry. 1978. Moisture Control Procedure and Products. *Technology and Conservation,* 3(1):38–42. **HP**

Cipriani, Carl. 1981. Justification for Specifying Protected Membrane Roofing Systems. Washington, DC: Office of General Services, Design and Construction, Division of Design. [2, 7]

Clark, E. J., P. G. Campbell, and G. Frohnsdorff. 1975. NBS Technical Note 883: Waterproofing Materials for Masonry. Gaithersburg, MD: National Bureau of Standards. [3, 4]

Council of Forest Industries. 1980 (Revised). *Western Red Cedar Shingles and Shakes: A Handbook of Good Practice.* Vancouver, B.C.: Council of Forest Industries of British Columbia. [6]

Commercial Renovation. 1987. The 1987 Premier Renovation Architects. *Commercial Renovation,* December: 24–44. **HP**

——— 1988. Sprayed on Roofing Prolongs Life. *Commercial Renovation,* April, 10(2):60. [5, 7]

Construction Specifications Institute, Construction Specifications Canada. 1983. *Masterformat.* Alexandria, VA: The Construction Specifications Institute. [8]

Copper Development Association, Inc. 1980. *Sheet Copper Applications.* New York: Copper Development Association. [8]

DeMuth, Jerry. 1986. White vs. Black: Does Membrane Color Matter? *Roofing Spec,* May, 14(5):25–28. [7]

Dhunjishah, Michael. 1986. The Roof Survey: Getting Back to Basics. In *Roofing '86.* Chicago, IL: National Roofing Contractors Association, pp. 11–16. [5, 7]

Doyle, Margret. 1987. Extending a Roof's Life. *Building Design and Construction,* June, 28(6):100–102. [7]

——— 1987. Keeping a Good Roof Down. *Building Design and Construction,* June, 28(6):108–111. [5, 7]

Dudley, Hubert T. 1987. Innovations in Vermiculite Roof Insulation Systems. *The Construction Specifier,* November, 40(11):62–65. [5]

Dupuis, Rene. 1987. Choosing the Right Roof Insulation System. *Exteriors,* Spring, 5(1):78–81. [5]

——— 1988. A Renewed BUR Technology Is Ready to Take on the 1990s. In *Handbook of Commercial Roofing Systems.* Cleveland, OH: Edgell Communications, Inc. pp. 16–23. [7]

Exteriors. 1986. Skyline: R-Values Questioned. *Exteriors,* Spring, 4(1):26–28. [7]

——— 1986. Roof Awareness. *Exteriors,* Autumn, 44. [1]

——— 1987. BUR Design Guide, *Exteriors,* Summer, 5(2):22–23. [7]

——— 1987. Certifying Consultants. *Exteriors,* Autumn, 41–42. [1]

——— 1987. Skyline: Roofing Innovations. *Exteriors,* Winter, 5(4):19–22. [7]

——— 1988. Skyline: Roofing Standards. *Exteriors,* Spring, 6(1):38–40. [7]

——— 1988. Skyline: Fiberboard Problems. *Exteriors,* Spring, 6(1):41–42. [5]

—— 1988. Skyline: Modified Bitumen Guide. *Exteriors,* Spring, 6(1):42–43. [7]

—— 1988. Portfolio: Florida Auditorium Roof Withstands Extreme Temperature Fluctuations. *Exteriors,* Spring, 6(1):68. [7]

Factory Mutual System. *Approval Guide.* Norwood, MA: Factory Mutual Systems. [5, 7]

—— *Loss Prevention Data Sheets.* Norwood, MA: Factory Mutual Systems. [5, 7]

Federal Construction Council. 1971 (August). Federal Construction Guide Specification Section 07-02, Metallic Waterproofing. Washington, DC: Federal Construction Council. [4]

Freund, Eric C., and Gary L. Olsen. 1985. Renovating Commercial Structures: A Primer. *The Construction Specifier,* July, 38(7):36–47. [1, 2, 3, 5, 6, 7, 8]

Fricklas, Dick. 1986. Structural vs. Architectural Metal Roofs. *Exteriors,* Summer, 8–8. [8]

—— 1986. Fire Codes. *Exteriors,* Autumn, 4(3):8. [7]

—— 1987. Better Products Help Prevent Roof Failures. *Building Design and Construction,* November, 28(11):60–63. [5, 7, 8]

—— 1988. Letter to the Editor. *Architecture,* January:10. [5, 7]

—— 1988. New Guidelines for Specifying Vapor Retarders. *Exteriors,* Spring, 6(1):8. [2]

General Services Administration. 1971 (January). Public Building Services Guide Specifications, Section 0712, Elastomeric Waterproofing System—Liquid Applied. United States General Services Administration. [4]

—— 1970 (February). Public Building Services Guide Specifications, Section 0751, Bituminous Roof Repair. United States General Services Administration. [6, 7]

—— 1970 (February). Public Building Services Guide Specifications, Section 0755, Repair of Built-Up Base Flashing. United States General Services Administration. [6, 7]

—— 1970 (February). Public Building Services Guide Specifications, Section 0762, Repair of Metal Roof Flashing. United States General Services Administration. [8]

—— 1975 (May). Public Building Services Guide Specifications, Section 07600, Sheet Metal, Flashings and Related Accessories, and amendments dated June 1975, and October 1977. United States General Services Administration. [8]

Gentry, Jim. 1986. Identifying Markets Key to Success in Reroofing. *Roofing Spec,* August, 14(8):21–24. [7]

Gordon, Douglas E. and M. Stephany Stubbs. 1988. Longevity and Single-Ply Roofing. *Architecture,* January: 95–99. [7]

Griffin, C. W. 1982. *Manual of Built-Up Roof Systems—Second Edition.* New York: McGraw Hill. [2, 4, 5, 6, 7, 8]

Grimmer, Anne E. 1984. A Glossary of Historic Masonry Deterioration Problems and Preservation Treatments. Washington, DC: Technical Preservation Services, U.S. Department of the Interior, Technical Report. **HP**

Harvey, John. 1972. *Conservation of Buildings*. London: Baker. **HP** [6, 8]

Hasan, Riaz. 1987. Eliminating Backout and Pullout of Roof Fasteners. *Exteriors*, Summer, 5(2):40–50. [5, 7]

Heineman, Paul. 1987. Coping With Membrane Roof Penetrations. *The Construction Specifier*, November, 40(11):36–43. [7, 8]

Henshell, Justin. 1987. Specifying Membrane Roofing: A Systematic Approach. *The Construction Specifier*, November, 40(11):102–108. [7]

Hoover, Stephen R. 1987. Wind Uplift on Roofs: One Perspective. *The Construction Specifier*, November, 40(11):76–79. [5, 7]

——— 1987. Single Ply Roofing: Exploring the Options. *The Construction Specifier*, December, 40(12):92–103. [7]

Huettenrauch, Clarence. 1985. Building Moisture and Roof System Destruction. *The Construction Specifier*, November, 39(11):15–16. [2, 5, 7]

——— 1987. Roof Supports and Roof Leaks: Avoiding the Inevitable. *The Construction Specifier*, July, 40(7):27–28. [7]

Indiana Limestone Institute of America, Inc. 1984–1985. *ILI Handbook*. Bedford IN: Indiana Limestone Institute of America, Inc. [3, 8]

Insall, Donald W. 1972. *The Care of Old Buildings Today: A Practical Guide*. London: Architectural Press. **HP** [6, 8]

Israel, Sheldon B. 1986. Reroofing: Principles for Success. *Exteriors*, Autumn, 4(3):50–53. [5, 7]

——— 1987. The Nuts and Bolts of Peripheral Roofing Components. *The Construction Specifier*, November, 40(11):44–49. [8]

——— 1987. Roof Renovation: A Viable Alternative. *Exteriors*, Autumn, 5(3):58–62. [5, 7, 8]

Jones, Larry. 1983. American "Thatch." *The Old House Journal* April: 59–60. **HP** [6]

Labine, Clem. 1983. How to Repair an Old Roof. *The Old House Journal* April:64–69. [6, 8]

LaCosse, Bob. 1986. Tech Talk: $75 Roof Proves a Real Bargain. *Roofing Spec*, August, 14(8):54. [7]

——— 1986. ASTM's New Terminology Gives Industry Chance to Re-Cover. *Roofing Spec*, November, 14(11):32–33. [1, 2, 5, 6, 7. 8]

——— 1987. End May Be in Sight for Aged R-Value Controversy. *Roofing Spec*, January, 15(1):17–20. [5]

——— 1987. The Air Force Manual Goes under the Knife. *Roofing Spec*, March, 15(3):15–18. [7]

Latta, J. K. 1972. Design of Weathertight Joints in Buildings: From Cracks, Movements, and Joints in Building. Record of Division of Building Research Science Seminar, Autumn. National Research Council of Canada. **HP**

The Lead Industries Association's *Lead Roofing and Flashing*. New York: The Lead Industries Association. [8]

Legatski, Leo A. 1987. Insulating Roof Decks with Cellular Concrete. *The Construction Specifier*, November, 40(11):56–61. [5]

Litvin, Albert. 1968. Portland Cement Association Research and Development Laboratories, Development Department Bulletin D137: Clear Coatings for Exposed Architectural Concrete. Portland Cement Association. (Reprinted from the May 1968, *Journal of the PCA Research and Development Laboratory*, 10(2):49–57.) [3]

Mack, Robert C. 1975. The Cleaning and Waterproof Coating of Masonry Buildings. Washington, DC: Technical Preservation Services, U.S. Department of the Interior, Preservation Brief No. 1. **HP** [3]

Maruca, Mary. 1984. 10 Most Common Restoration Blunders. *Historic Preservation*, October:13–17. **HP**

Miles, J. D., III. 1987. Tracking the Elusive Leak. *Roofing Spec*, March, 15(3):19–21. [7]

McCorkle, John. 1986. New Developments in Composite Roof Insulation. *Exteriors*, Summer, 4(2):40–42. [5]

Moreno, Elena Marcheso. 1987. Failures Short of Collapse. *Architecture*, July:91–94. [7]

Morton, W. Brown III. 1972. Field Procedures for Examining Humidity in Masonry Buildings. Association for Preservation Technology, Bulletin, VIII(2). **HP**

Nadel, Toby. 1986. The 25 Percent Solution. *Roofing Spec*, May, 14(5):35–36. [7]

National Concrete Masonry Association. 1981. NCMA—Tek 10A: Decorative Waterproofing of Concrete Masonry Walls. Herndon, VA: National Concrete Masonry Association. [3, 4]

——— 1973. NCMA—TEK 55: Waterproof Coatings for Concrete Masonry. Herndon VA: National Concrete Masonry Association. [3, 4]

National Roofing Contractors Association. 1985. *Quality Control in the Application of Built-Up Roofing*. Rosemont, IL: National Roofing Contractors Association. [4, 5, 7]

——— 1986. *The NRCA Roofing and Waterproofing Manual*. Rosemont, IL: National Roofing Contractors Association. [2, 3, 4, 5, 6, 7, 8]

——— 1986. *The NRCA Steep Roofing Manual*. Rosemont, IL: National Roofing Contractors Association. (Included as part of *The NRCA Roofing and Waterproofing Manual*. [6, 8]

——— 1986. Technology Harnessed to Seek Out Moisture. *Roofing '86:* 18–22. [5, 7]

———— 1986. Maintenance Programs Help Keep Roofs in Shape. *Roofing '86*:25–27. [5, 7]

———— 1986. *The NRCA Waterproofing Manual* Rosemont, IL: National Roofing Contractors Association. (Included as part of *The NRCA Roofing and Waterproofing Manual*.) [4]

———— *Roofing Materials Guide*. Rosemont, IL: National Roofing Contractors Association. [5, 7]

National Roofing Contractors Association and Asphalt Roofing Manufacturers Association. Revised 1988. *NRCA/ARMA Manual of Roof Maintenance and Repair*. Rosemont, IL: National Roofing Contractors Association and Asphalt Roofing Manufacturers Association. [5, 7, 8]

National Roofing Contractor's Association, Prospect Industries. 1986. Owners Want Answers to Protect Investment. In *Roofing '86*. Chicago IL: National Roofing Contractors Association, pp. 7–9. [5, 7]

National Slate Association. 1926 *Slate Roofs*. Vermont: National Slate Association. (Reprinted in 1977 by Vermont Structural Slate Co., Inc., Fair Haven, VT.) [6, 8]

National Tile Roofing Manufacturers Association, Inc. (NTRMA) 1987. *Installation Manual for Concrete Tile Roofing*. Los Angeles, CA: National Tile Roofing Manufacturers Association, Inc. [6]

National Trust For Historic Preservation. 1985. *All About Old Buildings—The Whole Preservation Catalog*. Washington, DC: Preservation Press. (This is an extensive reference work containing the names and address of many organizations active in the historic preservation field and lists of publication sources. Anyone facing a preservation problem should obtain this catalog as soon as possible. It will save much time in finding the right organization or data source.) **HP**

Naval Facilities Engineering Command. 1979 (May). Guide Specifications Section 07110, Membrane Waterproofing. Department of the Navy. [4]

———— 1983 (February). Guide Specifications Section 07140, Metallic Oxide Waterproofing. Department of the Navy. [4]

———— 1984 (July). Guide Specifications Section 07220, Roof Insulation. Department of the Navy. [5]

———— 1987 (March). Guide Specifications Section 07222, Tapered Roof Insulation. Department of the Navy. [5]

———— 1985 (January). Guide Specifications Section 07410, Preformed Metal Roofing and Siding. Department of the Navy. [8]

———— 1985 (October). Guide Specifications Section 07600, Flashing and Sheet Metal. Department of the Navy. [8]

Nimtz, Paul D. 1987. Comparing Structural and Architectural Metal Roofing, *Exteriors* Winter:46–49. [8]

The Old House Journal. 1983. Maintaining Your Roof: Inspection Checklist. *The Old House Journal* April:72–72. [5, 6, 7, 8]

——— 1983. Restoration Products News. *The Old House Journal* April:73–76. [6, 8]

Petersen, Wayne. 1987. Foamed Plastic Roof Insulation. *The Construction Specifier,* November, 40(11):66–74. [5]

Peterson, Arnold. 1987. Testing Single Ply's Weatherability. *The Construction Specifier,* November, 40(11):21–25. [7]

Phoenix Reprints. 1980. *Slate Roofing: Information You Need for Restoration and Repair.* Cincinnati, OH: Phoenix Reprints. [6, 8]

Poore, Patrica, 1983. Substitute Roofings—Credible Stand-Ins for Clay Tile, Slate, and Wood. *The Old House Journal* April:61–63. [6]

——— 1983. What Most Roofers Don't Tell You about Traditional and Historic Roofs. *The Old House Journal* April:56–58. [6, 8]

Porcher, Joel P., Jr. 1988. What You Should Keep in Mind When Specifying a Modified Bitumen. In *Handbook of Commercial Roofing Systems.* Cleveland OH: Edgell Communications, Inc., pp. 32–40. [7]

Portland Cement Association. 1982. Removing Stains and Cleaning Concrete Surfaces. Portland Cement Association. [3]

——— 1983. Permeability Tests of Masonry Walls. Portland Cement Association. [3]

——— 1986. Effects of Substances on Concrete and Guide to Protective Treatments. Portland Cement Association. [3, 4]

——— 1988. *Design and Control of Concrete Mixtures, Thirteenth Edition.* Portland Cement Association. [3]

——— Repairing Damp or Leaky Basements in Homes. Portland Cement Association. [3, 4]

Pruter, Walter F. 1987. The Advantages of Perlite Concrete Roof Deck Systems. *The Construction Specifier,* November, 40(11):52–55. [5]

Ramsey/Sleeper, The AIA Committee on Architectural Graphic Standards. 1981. *Architectural Graphic Standards, Seventh Edition.* New York: Wiley. [1, 2, 3, 4, 5, 6, 7, 8]

Red Cedar Shingle and Handsplit Shake Bureau. *Certi-Split Manual of Handsplit Red Cedar Shakes.* Red Cedar Shingle and Handsplit Shake Bureau. [6, 8]

Roofing Spec. 1987. On the Roof—Concrete Tiles Resemble Hand-Hewn Shakes. *Roofing Spec,* January:25–25. (*Roofing Spec's* name was changed in 1988 to *Professional Roofing.* [6]

——— 1987. BUR Repairs Made Easy with Torchable Modified Bitumen. *Roofing Spec,* March, 15(3):22–24. [7]

——— 1987. Techtalk: Drainage Requires Power of Positive Thinking. *Roofing Spec,* March, 15(3):46. [7]

Rossiter, Walter J., Jr. 1987. Nondestructive Methods for Inspecting Single Ply Roofing Membrane Seams. *The Construction Specifier* November, 40(11):92–100. [7]

——— 1988. Effect of Application Variables on Bond Strength of EPDM Seams. In *Handbook of Commercial Roofing Systems*. Cleveland OH: Edgell Communications, Inc., pp. 41–47. [7]

Roth, Michael. Siliconates–Silicone Resins–Silanes–Siloxanes. Germany. [3]

Russo, Michael. 1987. New Methods Needed to Measure Thermal Drift. *Exteriors,* Autumn, 5(3):68–74. [5]

——— 1987. A Critical Look at Thermoplastic Roofing. *Exteriors,* Winter, 5(4):24–27. [5, 7]

——— 1988. Survey: Architects Continue to Set the Trends. *Exteriors,* Spring, 6(1):56–66. [5, 7]

——— 1988. Going By the Numbers. *Exteriors,* Spring, 6(1):6. [6]

Single Ply Roofing Institute (SPRI). 1987. *Single Ply Roofing Systems: Guidelines for Retrofitting Existing Roof Systems*. Deerfield, IL: Single Ply Roofing Institute. [5, 7]

——— *Wind Design Guide for Ballasted Single Ply Roofing Systems*. Deerfield, IL: Single Ply Roofing Institute. [7]

Scharfe, Thomas R. 1987. Owners, Designers Discover Benefits of Metal Roofs. *Building Design and Construction,* June, 28(6):114–116. [8]

Sheet Metal and Air-Conditioning Contractors National Association, Inc. 1979. *Architectural Sheet Metal Manual*. Vienna, VA: SMACNA. [8]

Shell, Burt. 1988. Standing Seam Roof Continues Inrodes in Educational Market. *Metal Architecture,* 4(4):10–13. [8]

Shmedin, G. 1987. Bibliography for Slate Roofing. Association for Preservation Technology, Newsletter, 4(2):89. **HP**

Shinola Editions. 1980. *Tile Roofing*. Cincinnati, OH: Shinola Editions. [6, 8]

Single Ply Roofing Institute (SPRI). 1987. *Single Ply Roofing Systems: Guidelines for Retrofitting Existing Systems*. Single Ply Roofing Institute. [7]

Smith, Baird M. 1984. Moisture Problems in Historic Masonry Walls: Diagnosis and Treatment. Washington, DC: Technical Preservation Services, U.S. Department of the Interior, Technical Report. **HP**

Smith, George A. 1987. Fastener Corrosion to Be Judged by New Standard. *Roofing Spec,* January, 15(1):21–22. [5, 7]

Staehli, Alfred M. 1985. Historic Preservation: Where to Find the Facts. *The Construction Specifier,* July, 38(7):50–53. **HP**

Stahl, Frederick A. 1984. *A Guide to the Maintenance, Repair, and Alteration of Historic Buildings*. New York: Van Nostrand. **HP**

Stephenson, Fred. 1986. Clip Design Critical in Standing Seam Roofing. *Exteriors,* Spring, 4(1):60–65. [8]

Stover, Alan B. 1987. A Specifier's Guide to Construction Warranties. *The Construction Specifier,* November, 40(11):110–120. [7]

Sweetser, Sarah M. 1978. Roofing for Historic Buildings. Washington, DC: Technical

Preservation Services, U.S. Department of the Interior, Preservation Brief No. 4. **HP** [1, 2, 6, 8]

Thomas, Rose. 1987. Roofers Advise the Building Team How It Should Be Done. *Building Design and Construction,* June:92–94. [7]

Ting, Raymond. 1986. Metal Panel Behavior in Exterior Wall Design. *Exteriors* Autumn:65–69. [6, 8]

Tisthammer, Thomas. 1988. The Case for Specifying Sprayed Polyurethane Systems. *Handbook of Commercial Roofing Systems.* Cleveland OH: Edgell Communications, Inc. pp. 51–52. [5]

Tobiasson, Wayne. 1987. Vents and Vapor Retarders for Roofs. From *Proceedings of the Symposium on Air Infiltration, Ventilation, and Moisture Transfer.* Washington, DC: Building Thermal Envelope Coordinating Council. (Reprinted in the November 1987 *The Construction Specifier,* 40(11):80–90.) [2, 5, 6, 7]

Tobiasson, Wayne, and Marcus Harrington. *Vapor Drive Maps of the U.S.A..* Hanover, NJ: Cold Regions Research and Engineering Laboratory (CRREL), Corps of Engineers. [2]

Underwriters Laboratories. *Building Materials Directory—Class A, B, C: Fire and Wind Related Deck Assemblies.* Northbrook IL: Underwriters Laboratories. [5, 7]

——— *Fire Resistance Directory—Time/Temperature Constructions.* Northbrook, IL: Underwriters Laboratories. [5, 7]

United States Department of the Army. 1980. Engineers Guide Specifications, Military Construction, Section 07600, Sheet Metalwork, General. Office of the Chief of Engineers, Department of the Army. [8]

Vonier, Thomas. 1988. Metal Roofing Systems: Scraping Away the Impressions of the Past. In *1988 Handbook of Commercial Roofing Systems.* Cleveland OH: Edgell Communications, Inc., pp. 48–50. [6, 8]

Warford, Milan. 1988. What Aging Does to Roof Materials. *The Construction Specifier,* February, 41(2):29–30. [5, 7]

Warshaw, Ruth. 1986. Creative Design Options for Single-Ply Roofing. *Exteriors,* Spring, 4(1):42–49. [7]

West, Richard. 1987. Association Extends Invitation to Foam Contractors. *Roofing Spec,* January, 15(1):10. [5]

Williams, Jack B. 1986. Years of Work Went Into Making of *Quality Control. Roofing Spec,* August, 14(8):31–33. [7]

——— 1988. Designing the Total Integrated Roofing System. In *Handbook of Commercial Roofing Systems.* Cleveland OH: Edgell Communications, Inc. pp. 8–14. [5, 7]

Wilson, Forrest. 1984. *Building Materials Evaluation Handbook.* New York: Van Nostrand. [1, 2, 3, 6, 8]

Wiss, Janney, Elstner and Associates, Inc., D. W. Pfeifer and M. J. Scali. 1981. Report 244: Concrete Sealers for Protection of Bridge Structures. From *National*

Cooperative Highway Research Program. Washington, DC: Transportation Research Board, National Research Council. [3]

Wright, Gordon. 1987. Modified Bitumen's Market Share Growing. *Building Design and Construction,* June, 28(6):120–122. [7]

Index